THE NEURAL AND BEHAVIOURAL ORGANIZATION OF GOAL-DIRECTED MOVEMENTS

OXFORD PSYCHOLOGY SERIES

EDITORS
DONALD E. BROADBENT
JAMES L. MCGAUGH
NICHOLAS J. MACKINTOSH
MICHAEL I. POSNER
ENDEL TULVING
LAWRENCE WEISKRANTZ

The Neural and Behavioural Organization of Goal-Directed Movements

Marc Jeannerod

Professor of Physiology,
Laboratoire de Neuropsychologie
Expérimentale,
INSERM Unité 94,
Lyon, France

OXFORD PSYCHOLOGY SERIES NO. 15

CLARENDON PRESS · OXFORD
1988

Oxford University Press, Walton Street, Oxford OX2 6DP

Oxford   New York   Toronto
Delhi   Bombay   Calcutta   Madras   Karachi
Petaling Jaya   Singapore   Hong Kong   Tokyo
Nairobi   Dar es Salaam   Cape Town
Melbourne   Auckland

and associated companies in
Beirut   Berlin   Ibadan   Nicosia

Oxford is a trade mark of Oxford University Press

Published in the United States
by Oxford University Press, New York

British Library Cataloguing in Publication Data
Jeannerod, Marc
The neural and behavioural organization
of goal-directed movements.—(Oxford
psychology series; no. 15).
1. Goal (Psychology)
I. Title
150   BF503
ISBN 0–19–852117–0

Library of Congress Cataloging in Publication Data
Data available

Set by Promenade Graphics, Cheltenham

Printed in Great Britain by
St Edmundsbury Press Limited
Bury St Edmunds, Suffolk

Preface

This book deals with a problem central to the relations of brain and behaviour: the organization of goal-directed movements. The importance of movement, not only as a probe for studying brain mechanisms, but also as a clue to psychological functions, has been heralded by such eminent figures as H. von Helmholtz, R. S. Woodworth, C. S. Sherrington, N. Bernstein, P. Fitts, and many others. In addition, the study of movement has undergone, in the last 10 or 15 years, a complete review of its technical and conceptual background.

The use of behaving animals in neurophysiological laboratories has allowed direct investigation of the role of discrete neural mechanisms in controlling motor output. This type of preparation has—finally—fulfilled the ultimate requirement of any study dealing with motor control, the self-generation of spontaneous or trained goal-directed movements. Localized lesions of the brain areas involved or of their afferent of efferent pathways have now reproduced and documented in animals the classical human clinical syndromes which were, until recently, the only possible access to the study of motor deficits.

The renewal of interest in motor psychophysics, due in part to the availability of new techniques for recording movements, has drawn attention to the dynamic aspects of motor responses. A complete kinematic description of a movement can be regarded as a reasonable description of the neural commands responsible for this movement. It is therefore possible, by applying this reasoning to pathological or experimental conditions resulting from localized lesions, to infer the putative mechanisms for the generation of movements from descriptions of the movements themselves. This approach, sometimes referred to as the 'inverse kinematics' approach, is complementary to the traditional—and reverse—approach, which consisted of attempts to describe the complete set of neural events preceding and accompanying movements. The newer conception of motor control has the advantage of being directly transferable to other fields dealing with theoretical descriptions of motor systems and with their technological implications (e.g. artificial intelligence, robotics).

Finally, the introduction in the study of motor control, of experimental paradigms derived from cognitive psychology, has given substance and respectability to the abstract notion of internal representation. There is a definite trend towards acceptance that movements are not directly dependent upon sensory events, but that they are governed by representations which are built according to specific rules. Some of these rules are experimentally accessible and are being identified, not only in the representation of movement but also in other, related, representations for the properties of objects and their spatial relationships to the body.

This volume comprises six chapters, centered on the main aspects of the organization of movements. Many questions are raised (very few are answered) concerning the motor mechanisms and their implementation in goal-directed actions. In Chapter 1, dealing with motor programming, the central questions are: How detailed is a program? Is there a specific program for each movement, or is there a basic motor pattern that can be generalized to different movement types? Are complex movements organized on the same mode as simple ones?

Chapter 2 looks for order and hierarchy in the organization of motor patterns. How rigidly organized are co-ordinated actions like orienting or grasping? Is there a different neurological status for the proximal and distal components of the same action? Can the study of development of such motor patterns, and the study of the effects of brain lesions help in differentiating these components?

Chapter 3 examines the functions of visual feedback in relation to the execution of movements. How long does it take for visual feedback signals to influence the trajectory of a movement? Where in the trajectory visual feedback operates? What are the contributions of vision to movement accuracy and kinematics? And where do visual feedback signals arise?

Chapter 4 deals with the central problem of the directional coding of movements. How is the retinal map of visual space transformed into a body-centred map? How are relative eye and head positions signalled to the nervous system? Are there signals documenting the position of the body vector with respect to physical landmarks?

Chapter 5 is devoted to the role of position sense in controlling movements. Beyond the classical questions on the determinants of the sense of position of the limbs (What is the contribution of peripheral—proprioceptive—or central efferent mechanisms?), other questions arise: What is the contribution of vision to position sense? How are visual and proprioceptive signals integrated? How do they contribute to the representation of the mutual relationships of the body, the moving limb and the goal of the movement?

Finally, Chapter 6 explores the role of the posterior parietal cortex in the organization of movements directed toward extrapersonal space. Lesions occurring in these areas (in humans as well as in monkeys) produce a marked disorientation which affects not only the direction of the limb with respect to the target but also the kinematics of the movement. What is the nature of this deficit? To what extent does parietal lesion affect the representational level of movement generation? Can one conceive this type of lesion as disconnecting from each other the operators which compose the representation of the movements?

Most of these questions will remain open. Tentative answers will be offered, relying on a large body of experimental data from human and animal work, with particular emphasis on illustrative clinical cases.

Lyon   M. J.
1987

Acknowledgements

Most of the experimental work reported in this book, as well as many of the concepts which I have developed, are the fruit of a long-standing and enjoyable collaboration with Claude Prablanc. This work also owes much to my other collaborators at the Laboratoire de Neuropsychologie Expérimentale: B. Biguer, J. H. Courjon, S. Faugier-Grimaud, F. Michel, and M. T. Perenin. Oustanding technical assistance was provided during the writing stage by S. Bello, F. Girardet, P. Giroud, M. Rouvière, and C. Chonstan.

I wish to acknowledge the many friends and colleagues who have provided me with their encouragements and have participated in thoughtful discussions, particularly, M. A. Arbib, E. Bizzi, M. Brouchon, I. M. L. Donaldson, M. A. Goodale, A. Hein, C. von Hofsten, J. A. S. Kelso, D. N. Lee, C. MacKenzie, R. Marteniuk, D. E. Meyer, Ch. Phillips, R. Schmid, R. A. Schmidt, C. Trevarthen, P. Viviani, A. Wing, and L. Weiskrantz.

Finally, this work received constant financial support from INSERM, Paris and Université Claude Bernard, Lyon.

Contents

1. Models for the programming of goal-directed movements

Broadly defined, a motor program can be considered as part of the representation of an action, whether it be a simple movement or a more complex operation involving a succession of movements. Representation is a widely used, highly ambiguous terminology. On the one hand, it can be used to designate the mental content related to the goal or the consequences of a given action. On the other hand, it can be used outside the mentalist context, to designate 'covert' neural operations that are supposed to take place before an action begins (e.g. Saltzman 1979). Woodworth made clear this distinction between different levels of what he called 'intention': 'When I voluntarily start to walk, my intention is not of alternately moving my legs in a certain manner; my will is directed toward reaching a certain place. I am unable to describe . . . what movements my arms or legs are going to make; but I am able to state what result I design to accomplish.' (Woodworth 1906, p. 375). Such a representation of the action of walking as a goal-directed displacement of the subject relates to the conscious, or at least the cognitive aspect of that action. At the execution level, however, the same action has to be represented in a more detailed way, i.e. in terms of specific neural commands. It is therefore necessary to clearly distinguish at which level of representation an action is described.

1.1. Theoretical issues on the neural representation of movements

1.1.1. Cognitive models of motor programming

Different terminologies have been used to account for the notion that parameters for the execution of an action are represented both mentally and neurally. A first example is Head's concept of 'schema' (Head 1926). Although this concept was first used by its author to account for maintenance and regulation of posture, it was later developed by Bartlett as an internal model of the body in action; largely unconscious, and built from sensations and previous responses to external stimuli. [For a critical account, see Oldfield and Zangwill (1942). See also Schilder (1935) and Lashley (1951)]. More recently, the same concept was revived by Pew (1974) and Schmidt (1975). According to the latter author, the motor response schema arises as a relationship between information elements that are present within the context of executing a movement (e.g. initial

1

condition of the musculature, specification of the muscular commands, sensory consequences of the movement, and response outcome). Although this conception borrowed features from the previous ones, it departed from the original model in that it gave way to its mentalist content, and tended to include better-defined entities that could be accessible to empirical verification. Some of these entities, such as monitoring of initial state and specification of the commands, obviously have to exist before the movement is generated, while some others are consequences of it, as they appear through *reafference* generated by the movement itself, or through knowledge of the result of that movement. In this definition the schema also relates to the storage of information generated by movements and to motor learning. Accordingly, learning would result from interplay between elements that exist before the movement and those that result from it, until the actual outcome of the movement would correspond to the desired (represented) outcome (see Fig. 1.1 for a description of the Schmidt model).

Other models, like those of Arbib (1980) and Norman and Shallice (1980) are also constructed on the same principle. Norman and Shallice, for instance, assumed that specification of the components of actions was carried out 'by means of numerous memory schemas, some organized into hierarchical or sequential patterns, others in heterarchical or independent parallel (but co-operating) patterns, p. 5). Any given action sequence was represented by an organized set of schemas, with one—the source schema—serving as the highest order control and activating the other component-schemas for the individual movements of that action. When a given source-schema had been selected, component-schemas were controlled by 'horizontal' and 'vertical' processing threads. Horizontal threads determined the order of activation of the component-schemas and thus specified the structure for the desired action sequence, although vertical threads determined activation values for these schemas. Activation values involved attentional control, motivational factors, etc.

Marr's (1982) levels of functioning for information-processing devices represented the main theoretical framework for these 'top-down' models of motor representation. According to Marr, the highest level of the representation is the level of the 'computational theory', which defines the goal of the computation, its appropriateness, and the logic of the strategy by which it can be carried out. The intermediate level is that of the implementation of the computational theory into a representation of the input and output of the system and into an algorithm for transforming the input into output. The third level is the hardware implementation of the representation and the algorithm (see Marr 1982). Transferring Marr's concepts into 'motor' terms, one could suggest an analogy between his computational level and the 'cognitive' level of the motor representation (definition of the goal of the action and of the global strategy to achieve the

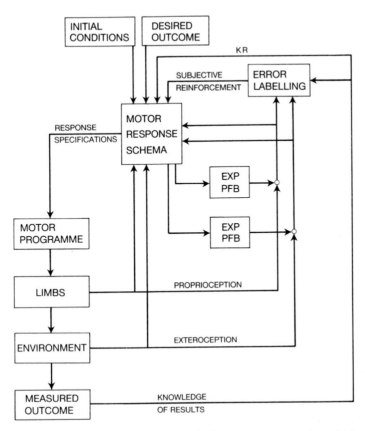

Fig. 1.1. The motor response schema in relation to events occurring within a trial. Abbreviations: KR = knowledge of results; EXP PFB = expected proprioceptive feedback; EXP EFT = expected exteroceptive feedback. (From Schmidt 1975, with permission.)

goal). The Marr algorithm level would correspond to the program for executing the movements, and the Marr hardware level would correspond to the neural network responsible for carrying out the motor commands.

A common feature of all these models is that they do not allow one to draw a clear separation between levels and do not offer a clear conception of their mutual relationships. In the case of the motor representation, a large degree of overlap seems to occur between 'cognitive' and 'motor' levels. Apparently, even the most elementary motor processes can be influenced (or 'penetrated') by specific cognitive states such as expectation, goal, or knowledge of the result. This notion of 'cognitive penetrability' (Pylyshyn 1980) suggests that the neural motor network (the Marr hardware level) might not be totally reducible to its own physical (anatomical and neurophysiological) properties.

1.1.2. Centralist and peripheralist theories

These points are important for understanding the nature of motor programs as they are conceived within the framework of the present chapter. Although the concept of the motor program in itself has never been seriously questioned, it has been attributed different degrees of explanatory value for the generation of movements. Motor theories stressing the role of a mental construct at the origin of movements (i.e. 'centralist', or program theories) have sometimes been disregarded because they were thought to favour the intrusion into motor processes of phenomena escaping pure physical causality. On the contrary, theories stressing a 'direct' transformation (i.e. without construction of a mental representation) of sensory input into motor output ('peripheralist' theories) represent a much more conservative alternative, whereby physiologically explainable (and cognitively impenetrable) mechanisms can account entirely for the process of producing a movement.

One consequence of this centralist versus peripheralist controversy is that programs have often been given a negative definition, such that a movement could be considered as resulting from a program only if it could be carried out uninfluenced by peripheral feedback. In fact, most classical definitions of the program (e.g. Keele 1968) are based on this default argument. This attitude may have reflected the need for early researchers in the field to escape the widely admitted notion that voluntary action had to result from sensory information in the same way as simpler (reflex) actions did. Demonstrating physiologically the existence of neural events independent from periphery and pre-existing execution of an action, may have been a necessary step before the notion of motor program could be accepted and a description of its internal structure undertaken (for a historical account see Jeannerod 1983).

This dichotomous thinking also reflects at the level of motor theories used to account for the more elementary aspects of action, for instance the execution of single movements. On the peripheralist side, the emphasis has been put on the role of sensory feedback, in clear continuity with engineering models that use peripheral feedback as a source of information about the response of a system. Adams (1971, 1976, 1977), for example, proposed that movement initiation was governed by a perceptual trace (a motor image) based on feedback from previous responses, and representing a reference of correctness for the movement. He considered movement execution as a closed-loop, self-regulating system that compensated deviation from this stored reference. Feedback generated during the movement itself was compared with the perceptual trace, and guided the limb to the target. In fact, Adams' theory is reminiscent of Sherrington's conception of movement initiation, according to which movements are initiated through the activation of stored motor images generated by reafferent sensory feedback.

In the centralist models, less importance has been given to the role of sensory signals. Held (1961), for instance, suggested that the internal reference accounting for completion of goal-directed movements should be derived from the program of these movements, as a corollary of the motor output, and independently from peripheral feedback. This model was based on the effects of visuomotor adaptation, in situations where the relation between the position of a visual target in space and the direction of movements toward that target was systematically modified by the use of laterally displacing prisms. Held and his colleagues proposed that, in order for the movements to become accurate under this condition, their program had to be modified from trial to trial and rebuilt according to the new visuomotor rules (see Held 1961). These authors based their hypothesis on the fact that active and passive movements during the prism exposure produced different effects. Adaptation (demonstrated by the existence of an after-effect after removal of the prisms) occurred only if the subjects made active movements, while passive movements (i.e. movements generating the same amount of visual and proprioceptive information as active movements but not resulting from a program) were ineffective in producing adaptation. It was therefore logical for Held to assume that programs at the origin of the active movements were also used as representations of the relations between the body and its environment, and that some comparison occurred between the effects of the movements and the corresponding representation. This theory clearly provides a basis for autonomy of the motor programs with respect to sensory feedback, yet does not exclude the role of reafferent sensory feedback in the comparison process (Fig. 1.2).

One of the ways of sorting out the respective contributions of peripheral and central mechanisms in motor programming is to study the effects of somatic deafferentation. Sherrington had hypothesized that if peripheral information from one limb were suppressed, movements would no longer be initiated with that limb. The interpretation given to the results of experiments involving deafferentation from peripheral feedback must be understood within this framework. Indeed, Mott and Sherrington (1895) showed that monkeys with a section of the dorsal roots corresponding to one arm were unable to correctly perform movements with their deafferented arm, particularly when movements involved the most distal segments of the limb (e.g. prehension). It was only when they were forced to use that limb that the monkeys could produce some awkward movements, but in normal situations the deafferented limb remained virtually unused. One animal was observed for up to 18 months with no evidence of recovery. Results obtained with the most recent monkey experiments, however (and

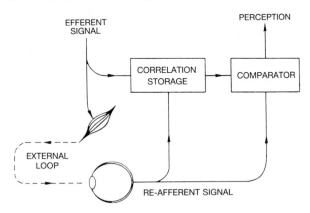

Fig. 1.2. Held's model, assumed to underly the consequences of sensorimotor rearrangement or disarrangement. The efferent signal from the motor program is split into a command signal for the muscles, and a copy signal, which is sent to the comparator. At this level the representation of the intended action is compared with the actual movement, which produces the reafferent signal. A passive movement would feed the comparator with reafferent signals only. (From Held 1961, with permission.)

also with human observations; for a complete review, see Chapter 5), lead to an interpretation of the role of peripheral feedback (particularly that of kinesthetic origin), quite different from that held by Sherrington. These experiments suggest that motor engrams can persist after deafferentation, in such a way that they can allow production of relatively complex movements. In addition, there are indications that motor engrams can be established independently of peripheral input. Berman and Berman (1973) and Taub *et al.* (1973) have reported that deafferentation of the upper limbs by bilateral dorsal rhizotomy, carried out on monkeys on the day of birth or even before birth, only slightly delayed the development of motor behaviour. It remains that behavioural production in such animals developed incompletely. They were permanently unable to perform normal finger movements (see Chapter 5).

Deafferentation experiments, through partly unconclusive, nevertheless clearly indicate that theories proposing an exclusive control of movement by feedback mechanisms are not supported by the available evidence. In fact, deafferentation data tend to favour the reverse conclusion attributing a primordial role to central processes. The main arguments of this theory were initially drawn from a clinical observation reported by Lashley (1917). This author described the case of a patient with a deafferented leg following a gunshot injury of the spinal cord. Anaesthesia was such that the patient was unable to detect the direction of any passive movement applied to his deafferented limb, nor the position of that limb after it had been passively displaced. Nevertheless the patient was able, in the absence of visual control, to move his leg on command in the proper direction, and to control the extent of his movements. Although he overestimated distance to be moved, his actual movements were found to be proportional to the distance that he had been asked to move. Lashley concluded from this study that accurate movement of a single joint was possible in the absence of any excitation from the moving organ. He later interpreted this result as evidence that movements must be entirely preset at the central level because, he thought, tactile, kinesthetic, and visual inputs triggered by movements had latencies too long to account for feedback regulation (Lashley 1951). The same argument, that some movements are too fast to allow time for feedback to influence performance, was used by Stetson and Bouman (1935), who pointed out that repetitive movements carried out by expert typists or pianists are essentially 'ballistic' and cannot be modified during their course.*

1.2. Identification of simple programming units: the pattern of muscular discharges related to reaching movements

Besides the theoretical issues raised by the notion of motor program, there is another, more pragmatic, line of research which aims at identifying simple program units and at describing their elementary components. These descriptions in fact have provided positive arguments for establishing the existence and the role of motor programs in the generation of movements.

One of the most widely accepted definitions of the program, that of Keele (1968), suggests that a motor program can be viewed as 'a set of muscle commands that are structured before a movement sequence begins, and that allows the entire sequence to be carried out' (p. 387). Thus, it appeared worthwhile to review in some detail the experimental data concerning muscular activity during discrete reaching movements. These studies will give information essential for understanding the nature and content of a program.

1.2.1. The triphasic EMG pattern

As early as 1895, Richer studied muscular contraction during fast alternating flexion–extension movements of the arm, by measuring changes in muscle shape on photographic films. Richer's main finding was that contraction of the agonist muscle ended before the end of the movement, so that the maximum extent of the movement coincided with relaxation of the

---

* Arguments as to a purely centralist theory of movement generation can also be drawn from the many reports that point to generation of complex motor patterns in completely deafferented preparations. A typical example is the alternating pattern of neuronal discharge related to locomotion in the isolated spinal cord in lower vertebrates (for a review, see Grillner 1985).

agonist muscle. Richer concluded that in such rapid movements, the limb was thrown like a ball by a sudden and momentary contraction of the agonist. By contrast, the antagonist was relaxed during contraction of the agonist (Richer 1895). However, the first detailed description of EMG discharges in relation to voluntary movements was not given until 1926, by K. Wachholder (Wachholder and Altenburger 1926). This author recorded agonist and antagonist muscle activity in normal subjects during flexion–extension movements of the arm. The limb position was also recorded by way of a mechanical device. The most consistent finding reported by Wachholder was the strict alternation between agonist and antagonist EMG activity: during movements performed at a moderate speed the initial burst of activity of the agonist muscle was followed by a burst of activity on the antagonist, appearing during the movement itself; finally a second burst of activity on the agonist was observed at the end of the movement. During fast movements, the same alternating pattern was observed. In addition, the bursts of activity of the agonist muscle tended to increase in duration when the movement amplitude increased, although the burst of the antagonist tended to become less and less marked (Wachholder and Altenburger 1926). Wachholder considered that this alternating pattern was determined centrally prior to movement onset. This point, he thought, was demonstrated by the fact that local anaesthesia of the antagonist muscle did not prevent appearance of the two bursts of activity of the agonist (see also Fessard 1926–7), a finding which was later replicated by Garland *et al.* (1972, see below).

Since Wachholder's work, the triphasic EMG pattern of simple voluntary movements has been amply confirmed (Terzuolo *et al.* 1973; Angel 1974; Hallet *et al.* 1975). The problem at this point is to determine the origin of each of the muscle bursts. The first burst of the agonist muscle (AG 1), which precedes movement onset is unquestionably of a central origin. Both amplitude and duration of AG 1 depend on movement amplitude, the larger the movement, the greater the amplitude and the longer the duration of the burst. Increase in burst amplitude therefore directly reflects gradation in force of contraction needed to produce larger limb displacements (Brown and Cooke 1981). This point is illustrated in Fig. 1.3(a), where EMG activities related to flexion movements of different amplitudes have been classified according to their rate of increase of force. The amplitude of AG 1 is clearly larger in movements where force increases more rapidly. In addition, in the same movements the peak force was shown to correlate with the amplitude of earlier events of the force trajectory [e.g. the second derivative of force d^2F/dt^2: see Fig. 1.3(b)]. These relations show that the amplitude of AG 1 reflects both the extent and the force of the movements (Gordon and Ghez 1984). It is interesting to note that the activity of pyramidal tract neurons has also been found, in monkeys trained to execute simple movements, to correlate with force, or

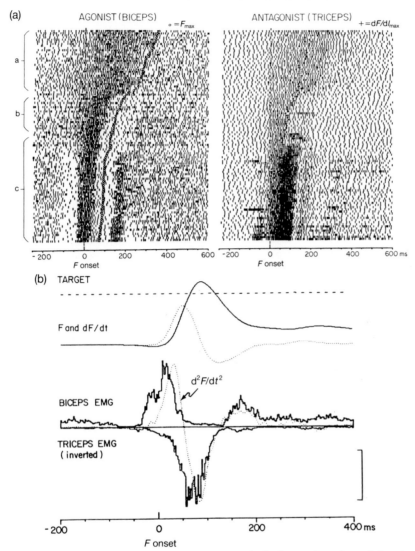

Fig. 1.3. (a) Change of the first agonist burst (AG 1) as a function of the rise time of force pulses during forearm flexion in normal humans. Density of small bars indicates degree of contraction of biceps (agonist) and triceps (antagonist) muscles over repeated trials. Trials were ordered as a function of decreasing rise time of force. *F* onset: onset of the curve d*F*/d*t*. Small circles: peak of d*F*/d*t*. Note higher and sharper EMG bursting when the time between onset of force and peak force shortens.

(b) Dynamic and EMG patterns of forearm flexion, averaged from 16 trials. Note the very clear triphasic EMG pattern with onset of AG 1 preceding the onset of force and onset of ANT overlapping with AG 1. (From Gordon and Ghez 1984, with permission.)

rate of change in force, rather than with other parameters of the movements, such as joint position (see Evarts 1968).

Increase in burst duration, however, is more difficult to understand, particularly if one considers that the time to maximum amplitude of AG 1 is relatively invariant for different levels of force (Freund and Büdingen 1978). In fact, AG 1 duration seems to be constant (around 70 ms) for movements of small amplitude, up to about 20° degrees (Hallett and Marsden 1979; Brown and Cooke 1984), although it sharply increases (up to about 200 ms) for larger movements (Berardelli *et al.* 1984; Brown and Cooke 1984; see also Wadman *et al.* 1979). Increase in the duration of AG 1 with movement duration (and velocity; Lestienne 1979) might be the case only for simple movements like flexion–extension, the duration of which increases with amplitude beyond a certain range. It is interesting to note that increase of AG 1 duration in movements of increasing amplitude was found to persist in one subject with functionally deafferented arms, hence demonstrating the purely central nature of this component (Cooke *et al.* 1985).

1.2.2. Role and origin of the antagonist burst

The origin of the following two bursts of the triphasic EMG—the second on the antagonist (ANT) and the third on the agonist muscle (AG 2)—is less easily determined. Ghez and Martin (1982) have argued from their experiments on cats that the EMG pattern during rapid force changes at a single joint is different, depending on whether a displacement occurs or not. In isometric force changes (i.e. without displacement), only AG 1 is present, but in non-isometric changes (with displacement), the typical three-burst pattern is observed. From these observations, it might be concluded that ANT and AG 2 are mere responses to movement, a hypothesis reinforced by the fact that, according to Ghez and Martin (1982), the timing of ANT with respect to limb displacement is the same in active and passively imposed movements. However, the antagonist burst has been shown to be often initiated prior to the termination of AG 1 (Hallett *et al.* 1975), sometimes even before a significant displacement of the limb segment has occurred (Angel 1977). In addition, both Hallett *et al.* (1975) and Angel (1977) have shown that ANT is still present in cases where the movement resulting from activity of the agonist muscle is actually prevented. According to Angel (1977), the antagonist burst decreases in the condition where the movement is blocked just prior to initiation, therefore suggesting some influence of proprioceptive loops in its generation. The same finding was reported by Denier van der Gon and Wadman (1977) and Wadman *et al.* (1979).

The function of the antagonist burst is still a matter of discussion. ANT might represent a central 'braking' strategy, a mechanism achieving final accuracy of the movement. An experiment of Waters and Strick (1981) tends to demonstrate this point. Subjects had to perform flexion–extension movements of the wrist by moving a handle, in order to match the position of a target on a screen. On some trials, however, the instruction was not to match the target, but to displace the handle as fast as possible against a stop. In the target condition, an antagonist burst was clearly present, the amplitude of which seemed to correlate with the velocity of the movement. By contrast, in the stop condition, no EMG activity was recorded on the antagonist muscle, even though movements had the same amplitude and velocity as in the target condition (Waters and Strick 1981). In addition, Marsden *et al.* (1983) have shown that ANT amplitude and timing are both influenced by the extent and velocity of the movements. Movements made through large angles require less antagonist activity than those made through small angles at the same velocity. ANT timing also changes as a function of both velocity and extent, with the burst occurring earlier in small fast movements than in large slow ones. In the former case, ANT started soon after the onset of AG 1. These results would fit into the interpretation of Gordon and Ghez (1984), who proposed that the role of activity in the antagonist muscle might be to compensate for non-linearities in muscle properties, and to allow linear scaling of forces.

The fact that ANT is not a stereotyped phenomenon, and can be influenced by the subject's strategy (an example of 'cognitive penetrability'), strongly indicates that this component belongs to the program of accurate movements, the function being to decelerate and stop the movements at the target, and to produce linear trajectories. The above-mentioned results of Marsden *et al.* (1983), together with results from experimental situations where movements of segments with a low inertial load were studied, tend to support this hypothesis. Meinck *et al.* (1984) for instance, have shown that accurate isotonic finger flexion, even for an amplitude as large as 30°, could be achieved (in about half the subjects) without activity of the antagonist muscle. In these subjects, AG 2 was also absent in 70 per cent of trials. This finding fits Lestienne's contention that an isotonic movement can be braked passively if the agonist force does not exceed the passive viscoelastic tension developed by the muscles involved (Lestienne 1979). Meinck *et al.* (1984) have also shown that an antagonist burst of activity consistently occurred during accurate isometric finger flexion (i.e. without displacement) if the instruction to the subject was to reach the target by pressing on a gauge as rapidly as possible. By contrast, ANT disappeared if during the same task, the subject was instructed to hold the force after the movement. Meinck *et al.* (1984) concluded that occurrence of ANT and AG 2 during voluntary movements is not part of an invariable motor subroutine, but depends on the inertial properties of the limb to be moved and on the task to be achieved.

Finally, an experiment of Garland *et al.* (1972), replicating the observation made by Wachholder and Altenburger (1926), illustrates the central nature of AG 2. An EMG of the triceps was recorded in normal subjects during arm extension movements. Anaesthetic block of afferents from the biceps (the antagonist muscle) was produced by local injection of lidocaine. A clear two-burst pattern of muscular activity nevertheless persisted in the triceps during voluntary elbow extension. Garland *et al.* (1972) concluded that the inhibitory pathway from antagonist to agonist had no influence in modulating motor outflow during rapid voluntary movements.

The central origin of the last two bursts in the EMG sequence is also strongly suggested by the fact that both ANT and AG 2 can be recorded in subjects with peripheral deafferentation of the moving segments. A normal triphasic EMG pattern was observed during flexion–extension movements by Hallett *et al.* (1975) and by Rothwell *et al.* (1982) in patients suffering peripheral sensory neuropathy, with loss of sensation at the hand and forearm levels (see also Chapter 5). A similar patient was observed by Forget and Lamarre. On a first examination, this patient had difficulty in performing accurate rapid flexion elbow movements: during such movements AG 1 and AG 2 were present, but no antagonist activity was observed (Forget and Lamarre 1981). One year later, however, the patient's motor performance was much improved, despite the fact that he was still functionally totally deafferented. Along with the more accurate motor performance, an antagonist burst was now present, with a constant duration of about 100 ms and ending precisely at the time of maximal deceleration of the movement (Forget and Lamarre 1982). These data obtained in human patients seem to contradict those obtained by Terzuolo *et al.* (1974) in deafferented monkeys. In such animals antagonist activity disappears during ballistically induced flexion movements at the elbow level. However, in view of the previously mentioned data on dependence of ANT on type of movement and type of task, one cannot think of Terzuolo *et al.*'s findings as being in direct contradiction with the hypothesis of central origin of this component. In addition, deafferented monkeys seem to have an increased spontaneous muscle activity in both agonist and antagonist muscles at rest, and therefore an antagonist burst would not be necessary for braking the movements.

It is interesting to compare the above results obtained from permanently deafferented patients with those obtained in normal subjects transiently deafferented by local anaesthesia or ischaemic block. Sanes and Jennings (1984) have studied EMG patterns during active or passively induced flexion–extension wrist movements directed at visual targets in subjects during ischaemic block above the elbow. Figure 1.4 shows agonist and antagonist EMG of voluntary movements executed before and during the block. Before the block, when cutaneous and kinaesthetic sensations were still intact, rapid voluntary flexion movements were accompanied by the

NORMAL   **ISCHEMIC**

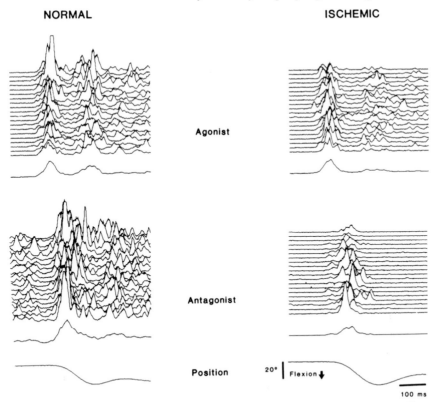

Fig. 1.4. Independence of the triphasic EMG pattern from somatosensory afferents. Agonist and antagonist EMG activity was recorded during 20 wrist flexion
movements in one normal subject. Integrated EMG activity is displayed for each
trial (from bottom to top) and also as an averaged curve for the 20 trials (separate
tracing). Position of the wrist is also indicated at the bottom of the figure.

In the normal condition, the triphasic EMG pattern is clearly seen both on individual and on averaged tracings. Under ischaemic block the same pattern is
observed, though EMG responses are globally smaller. (From Sanes and Jennings
1984, with permission.)

typical triphasic EMG pattern. This pattern was observed whether the
movement was isotonic or isometric. Following ischaemic deafferentation
of the forearm, the EMG patterns relating to active isotonic or isometric
flexion movements were strikingly similar to those obtained in the normal
condition. Timing of the bursts relative to the movement or relative to each
other was unchanged, though the overall amplitude of the bursts was
slightly reduced. By contrast the EMG pattern related to passively induced
flexion differed in the two situations. Before ischaemic block a typical
stretch response was recorded in the extensor muscle. This response
disappeared during ischaemia (Sanes and Jennings 1984).

1.3. The impulse model of motor programming

Demonstration of the impulse nature of central motor commands related to production of ballistic movements has stimulated a considerable amount of theoretical interest and experimental work. The idea that command impulses might be sufficient for generating the complete extent of a movement, with no or little intervention of feedback control loops (the impulse theory) in fact also originated from earlier sources. Lashley's (1951) statement that movements were too fast to owe much to peripheral mechanisms was highly influential. In addition, a long-standing interest was raised by Woodworth's (1899) demonstration that reaching movements were composed of several submovements which he thought were produced by discrete motor impulses. Later, this theory was adopted in the cybernetic context, particularly when Crossmann and Goodeve (1963) suggested that movement accuracy was achieved by iterative correction pulses of decreasing amplitude (see Chapter 2).

1.3.1. The force–time function and the idea of a generalized motor program

Theoretical implications of the impulse theory of motor programming models have been extensively discussed in two important papers by Schmidt *et al.* (1979) and Meyer *et al.* (1982). According to Schmidt and his colleagues (Schmidt *et al.* 1979), the motor system would generate muscular impulses for which force, duration, and onset can be controlled centrally. Impulses of this sort can be characterized as acceleration–time, or force–time, functions. It is assumed that force–time functions tend to remain invariant for movements having different task demands. Force and/ or duration parameters of a given function can be scaled to each other without affecting the mathematical form of the function. For instance, force can be increased within a constant timing (hence producing movements of larger amplitude with the same duration), or the reverse. In addition to scaling of the impulse generating mechanism, Schmidt and his colleagues postulated an in-built variability of muscular impulses. They assumed that variability in force of impulses is proportional to the amount of force produced, and variability in their duration, proportional to the impulse duration.

These relations allow interesting predictions. For instance, making movements of a given amplitude faster should result in reduction of variability in impulse duration. On the other hand, making movements of a given duration larger should make variability proportional to force. The latter prediction yields to an inverse relation between speed and accuracy of movement; the shorter the movement duration, the larger the error.

Schmidt *et al.* (1979) actually found such a linear function in measuring errors in amplitude in very fast pointing movements (less than 200 ms in duration). This point will be discussed further (see Chapter 3).

The simple impulse variability model of Schmidt and his co-workers takes no account of the fact that even in ballistic movements impulses are not exclusively sent to agonist muscles, but also to antagonist muscles. Accordingly, any model of impulse generation should deal equally with acceleration and deceleration of movements. Meyer *et al.* (1982) included in their model, which shares the same basic premises as that of Schmidt *et al.*, a strict symmetry of acceleration and deceleration phases of the force–time function. Yet, movements with symmetrical acceleration/deceleration profiles may represent a rather artificial type. As will be emphasized below, most movements executed in normal conditions are far from being symmetrical, in relation to both the size and duration of the acceleration and deceleration phases.

An interesting problem arises with regularity of force–time functions. According to the Meyer *et al.* model the prototypical force–time function can be rescaled each time either parameter varies, so that force–time functions corresponding to a given task will keep the same basic shape. This property of rescalability implies that force and time parameters can vary independently of each other (force–time decomposition). Conservation of the 'shape' of a force–time function implies stability in relative movement time of the different kinematic landmarks such as peak acceleration, zero crossing, etc. This mode of impulse constancy is achieved in certain types of target-aimed movements where movement duration tends to remain constant for different values of force (e.g. Freund and Büdingen 1978; Jeannerod 1984; see also Milner 1986) (Fig. 1.5(a)). This type of movement was considered by Ghez (1979) as resulting from a single command impulse in which amplitude varies, but where duration is fixed (the pulse height control model). More commonly, shape constancy is achieved by rescaling kinematic landmarks in relation to a varying movement duration. An example is saccadic eye movements. In this case, where the duration of the movement clearly correlates with its amplitude, the time to reach maximum force is also proportional to movement duration (Robinson 1964; Van Opstal *et al.* 1985) (Fig. 1.5(b)).

There are other situations, however, where the principle of shape constancy is clearly violated. For instance, in conditions where the timing of a movement can be controlled voluntarily, force seems to be rescaled independently of duration (time to maximum acceleration remains invariant). This is achieved at the expense of shape constancy, since the other landmarks such as zero crossing and peak deceleration are displaced as a function of movement duration (Zelaznick *et al.* 1986). Another obvious violation of shape constancy is represented by movements requiring a high accuracy and where the duration of the deceleration phase varies as a function

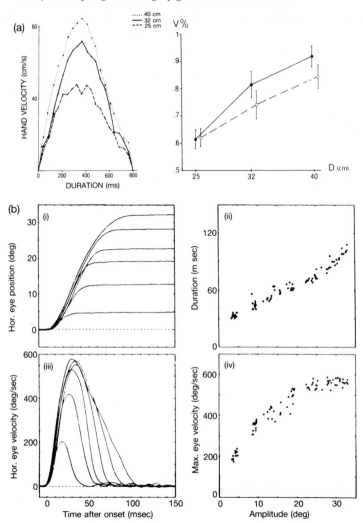

Fig. 1.5. Invariance of force–time functions in scaled movements.

(a) Hand movements during reaching for objects located at 25, 32, or 40 cm of body. Velocity profiles as a function of time for three individual movements from the same subject are represented. Note constant duration and increasing peak velocity when movement amplitude increases. Diagram on the right: relationship of peak velocity (V) to movement amplitude (D) in reaching movements from three subjects. Solid line: no visual feedback available from the moving hand. Dashed line: no visual information available, the target being turned off at movement onset. (From Jeannerod 1984.)

(b) Saccadic eye movements in the horizontal plane in one normal subject. (i) eye position; (ii) eye velocity; (iii) relationship of saccade duration to saccade amplitude; (iv) relationship of peak velocity to amplitude. (From Van Opstal *et al.* 1985, with permission.)

of total movement duration, although the duration of the acceleration phase remains invariant.

This point is illustrated by an experiment of Marteniuk *et al.* (1987) where reaching movements in different conditions of task constraint were compared. A constraint is defined here as a variable that limits the manner in which movement control occurs. As an example, the target size can be considered as a movement constraint in that movements aimed at small targets have velocity profiles distinctly different from those produced in attaining larger targets (Soechting 1984). In the experiment of Marteniuk *et al.*, the goal of reaching movements, as well as the required movement extent and accuracy, were systematically varied. Subjects were asked either to hit a target (2 cm or 4 cm in diameter) using the index finger, or to grasp a 1-cm thick disc (2 cm or 4 cm in diameter) between the thumb and index finger. Both targets and discs were placed at a distance of 20 cm or 40 cm from the resting position of the right hand. Movement trajectories were recorded in three dimensions with a high sampling rate (200 Hz), allowing measurement of movement time and reconstruction of velocity profiles.

Fitts' law was found to apply in both types of movement. Movement time was longer during grasping than during hitting, but in both cases it increased as a function of difficulty of the task (as quantified by combining the target/disc width and the movement amplitude) (Fig. 1.6). Both types of movement had similar values of maximum velocity, and in both cases maximum velocity was higher with increasing target/disc width and with increasing movement amplitude. The most interesting difference between the two types of movements was the repartition of movement time into acceleration and deceleration. The peak velocity during hitting and during grasping was reached at about the same time, and the longer duration of grasping movements was due to a higher percentage of total movement time spent in the deceleration phase (Fig. 1.7).

Marteniuk *et al.* (1987) also explored further the determinants of the movement kinematics by comparing grasping movements in two different situations. In both, the subject had to grasp the disc as in the previous experiment; in one situation he had to throw it in a large container, and in the other one he had to tightly fit it in a small container. Only the reaching parts of the movements were analysed. This comparison was undertaken with the aim of exploring whether the context within which a movement is executed can affect its characteristics. Indeed movement time was found to be longer for grasping movements executed prior to fitting than for movements executed prior to throwing. This lengthening in movement time was due to lengthening of the deceleration phase (Fig. 1.8).

These data clearly indicate that planning and control of trajectories do not occur through the use of a general and abstract representation, which would adjust movements for different task conditions by simply scaling a

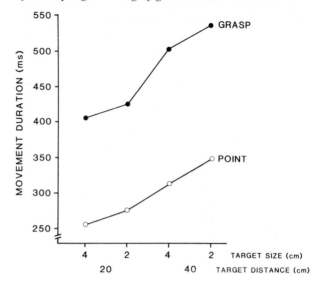

Fig. 1.6.   Relationship of movement duration to task difficulty in reaching. Total movement duration has been measured in the tasks of pointing at targets and grasping discs of the same size. Target/disc sizes and distances from the subject were combined systematically in order to vary the 'difficulty' of the task (for example, a trial involving a large target at a short distance is considered an 'easy' task, while a trial with a small target at a long distance is considered 'difficult'). In both pointing and grasping tasks, movement duration increases monotonically with task difficulty. (Data from Marteniuk *et al.* 1987.)

basic velocity profile. Instead, higher order factors such as the goal, the context, (and also probably the knowledge of the result) of the action seem to be able to influence not only duration and velocity, but also the intrinsic kinematic structure of the movements. Impulse form constancy therefore only holds in a broad sense and cannot be taken as a proof for independence of force and time generation mechanisms. These considerations make somewhat illusory the idea, implicit in the impulse variability model, of a generalized motor program representing the elementary unit for any type of action.

1.3.2. Implications of the impulse model for the production of accurate reaching movements

In order to achieve a desired final position, impulse-like motor commands applied to a musculoskeletal system have to take into account, not only the distance to be covered to reach the target, but also the mass of the system and the biomechanical properties of its muscles.

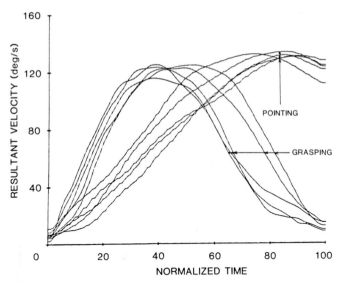

Fig. 1.7. Velocity profiles of pointing and grasping movements. The grasping movement involves an acceleration phase (up to the peak velocity value) shorter than the deceleration phase. The reverse is true for pointing. Note that during pointing the deceleration is interrupted by the fact that the subject hits the target at a relatively high velocity, a strategy which is not compatible with grasping. Also, note that movement time was normalized. In fact, as clearly shown in Fig. 1.6, pointing movements have a much shorter duration than grasping movements. Movements recorded in three dimensions with a high-rate sampling video technique (WATSMART). (From Marteniuk *et al.* 1987.)

(a) Length–tension muscle properties and the equilibrium point hypothesis

It has been known for a long time that muscles have predictable characteristics which relate the static force that they can produce (their strength) to the angle of the joint to which they are attached. Accordingly, each muscle can be characterized by a length–tension curve, where its strength is an increasing function of the angle of the joint (the strength of the muscle increases with its length) (Matthews 1959; Astratyan and Feldman 1965; Feldman 1966). This fundamental property of muscles will obviously determine the amplitude and the duration of the command impulses to be applied to the muscles, because the force needed to produce rotation of a joint by a given angle will vary according to the resting position of that joint before the movement. These static muscle properties are partly controlled by the central nervous system itself, which can change the degree of activation of the muscle. Finally, there is a possibility for feedback control to regulate the length–tension curve, via a change of muscle stiffness.

Another characteristic of the musculoskeletal system is the load against

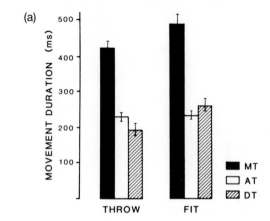

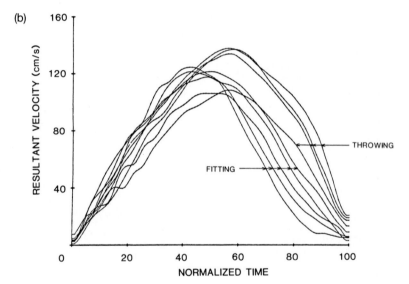

Fig. 1.8.   Influence of task constraints on the temporal pattern of reaching movements.

(a) Movements have been recorded as in Fig. 1.7 during the action of grasping. On some trials subjects had to grasp the disc and throw it away. On other trials, they had to grasp the disc and fit it in a small box. These task constraints influenced both total movement duration (MT), which was longer in the fit condition, and the velocity profiles. In the throw condition acceleration time (AT) was longer than acceleration time (DT), although the reverse was observed in the fit condition. DT was about 80 ms longer in the fit condition than in the throw condition.

(b) Representative velocity profiles in the two conditions. Note that time has been normalized for enabling comparison. (Data from Marteniuk *et al.* 1987.)

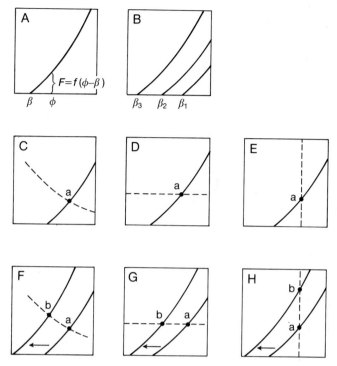

Fig. 1.9. Hypothesis of equilibrium point. A. An invariant characteristic (tension F vs angle φ) of a muscle subserved by servoregulatory mechanisms. β: threshold of tonic activation of the muscle. B. Invariant characteristics corresponding to different values of β. C–E, Equilibrium point (a) as a point of intersection of muscle and load characteristics. Dashed lines: characteristics of isometric E, isotonic D, and intermediate C load. F–H. Changes of force H, angle G or both F caused by a shift of equilibrium point (a → b) under the respective load conditions (dashed lines). (From Feldman 1981, with permission.)

which the muscle contracts. This load is caused by the mass of the system, the tension of antagonist muscles, and viscoelastic properties of surrounding tissues. Both the load and the muscle characteristics interact to produce an equilibrium point where the system is stationary. If the degree of muscle activation is changed by the central nervous system, muscle characteristics will shift to a new length–tension curve. This will determine a new equilibrium point of the muscle-load system, and movement will occur as a consequence of the shift from one equilibrium point to another until a new stationary state has been reached (Fig. 1.9).

Feldman's description of the predictable muscle characteristics can be modelled by a simple mechanical analogy, where muscles are compared to adjustable springs. Force produced by a spring is related to its degree of stretch (its length); adjusting its resting length or its stiffness changes the relation between degree of stretch and force. Accordingly, a limb segment can be modelled as a pair of adjustable springs acting across the joint in the agonist–antagonist configuration against the mass of the segment (the mass–spring model, Feldman 1966). This model has had a broad impact on the conception of motor programming. Specifically, it has led to a simplification of the notion of program by limiting its content to the control of a few parameters, like muscle stiffness. Provided the position of the target is accurately computed (e.g. on the basis of visual cues) and the muscle stiffness is changed accordingly, the desired position of the limb will be reached automatically. This conception has a number of important potential consequences:

(a) programming arrest of the movement should become unnecessary;

(b) the change in limb position should be achieved without the contribution of feedback control;

(c) no prior 'knowledge' of the state of the muscle–load system should be necessary at the program level before the movement is initiated; and

(d) the movement trajectory should not need to be controlled.

One of the major simplifications of the programming procedure introduced by the mass–spring model is that the desired position of the limb may be specified independently of initial conditions. This is an old problem in psychology (the motor equivalence problem): namely, how a particular single goal can be reached in several different ways and from several different starting points. The various motor patterns produced by an individual to reach the same goal can be considered as functionally equivalent, even though they may widely differ from each other. Particular emphasis has been given to this problem by authors working in the field of motor control of speech production. One of the main questions that speech production theorists have to answer is, how do articulators (which determine the configuration of the vocal tract) come as close to reaching the same final position as they do, for the production of a given phoneme? One possible answer is that each phoneme would be produced by an invariant and specific motor command. In fact, analysis of EMG articulator patterns during speech shows a relatively large variability for a given phoneme, according to which phoneme precedes or follows it, to speaking rate, etc. An alternative hypothesis has been proposed to account for this difficulty, by assuming that the command is generated, not to produce a certain utterance, but for the vocal tract to adopt a certain target position. This hypothesis (Mac-Neilage 1970) implies internalization (or representation) of a space coordinate system and specification of targets within this internalized space. It seems appealing within the context of language production, because it allows the conception of the motor commands responsible for reaching the target phoneme as context independent and free from feedback control, an obvious advantage for rapid motor sequencing. There are limitations to this theory, however, that will become apparent in a later section of this chapter.

(b) Evidence and limitations for the equilibrium point hypothesis

In theory, motor programs could thus be reduced to a simple coding of final limb position. This hypothesis has led Bizzi and his co-workers to a series of interesting experiments in monkeys. Basically, these experiments involved application of unexpected transient loads at the onset of trained movements directed at visual targets. According to the theory, this perturbation should not prevent the animal from reaching the final desired position, because, after the load has been removed, the moving segment should stop at the equilibrium point specified by the degree of activation of the agonist and antagonist muscles. In a first set of experiments dealing with eye and head coordinated movements in vestibulectomized monkeys a constant torque disturbance was applied at the beginning and throughout the centrally initiated movement. This produced an undershoot of the head movement with respect to the target, followed, after cessation of the torque, by immediate readjustment, driving the head at the target position (Fig. 1.10(a)). Because the visual target was turned off after movement onset and vestibular function was abolished by vestibulectomy, the readjustment could not have been due to correction based on visual or vestibular input. When the load was applied only during the dynamic phase of the movement (i.e. as a brief perturbation instead of a constant torque), the amount of force developed by the head was increased (presumably via an increased muscle spindle discharge during tension of the muscle) and the head overshot the target position. After cessation of the load, however, the desired final position was correctly achieved (Bizzi *et al.* 1976) (Fig. 10(c)). These results seem to indicate that central programs for simple movements are not influenced by the amount of proprioceptive information generated during the movement. In order to bring a direct proof as to this point, the same experiments were repeated in monkeys in which the head–neck system had been deafferented from any sensory control (open-loop condition), by destruction of the vestibular apparatus and section of the dorsal roots corresponding to the neck muscles. Indeed, after deafferentation, the animals could still respond accurately to the visual targets, including in the loaded situations (Bizzi *et al.* 1976, 1978) (Fig. 1.10(b, d)).

Further experiments exploring acquisition of a visual target with arm pointing movements, instead of head movements, by deafferented monkeys were also reported by Bizzi's group. Monkeys were trained to reach the location of a target zone (10–15° wide) by flexion–extension elbow movements. Vision of the arm was prevented during the experiment. The manipulandum to which the monkey's arm was attached could be unexpectedly loaded either during the movement itself or some 150–200 ms prior to movement initiation. In the latter case, the load produced a passive

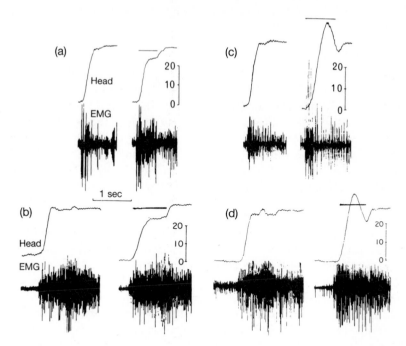

Fig. 1.10.   Head movements executed under loaded conditions by deafferented monkeys.

(a) Vestibulectomized animal. On the left, normal, unloaded head movement, directed at a visual target. Right, movement at the same target under condition of constant torque (continuous line on top of record). Note undershoot followed by resetting to intended final position after unloading.

(b) Vestibulectomized and deafferented (section of C1–T3 dorsal roots) animal. Head movements executed toward visual target in the unloaded condition (left) and the constant torque condition (right). Resetting of position after unloading occurs in spite of deafferentation.

(c) Vestibulectomized animal. Head movement toward a visual target in the unloaded condition (left) and under condition of a dynamic perturbation applied during the movement. Note overshoot and resetting of intended final position.

(d) Vestibulectomized and deafferented animal, as in (b). Records on left and right, same conditions as in (c). Note persistence of resetting in spite of deafferentation. EMG: electromyogram of splenius capitis. (From Bizzi *et al.* 1976, with permission.)

displacement of the forearm in the direction of the load, i.e. toward the target or away from it. Intact monkeys had no problem reaching the target zone when their forearm was passively displaced by rotation at the elbow joint prior to movement onset. A high degree of correlation was observed between final position of the hand and target position. Following deafferentation of their arm by section of the corresponding dorsal roots, the monkeys performed in an essentially similar way (correlation coefficients were within the same range as before deafferentation), in spite of being totally unaware of the perturbation when it occurred (Polit and Bizzi 1978, 1979).

In contrast with normal monkeys, however, deafferented monkeys could only compensate for perturbations occurring at the elbow joint. They were unable to compensate for perturbations affecting the canonical position of their arm, for instance when the point of rotation of the elbow was displaced away from their body. Inability to reach the target position in that condition indicates that determination of the final hand position in fact also requires encoding of the position of limb segments with respect to the body prior to movement execution, and underscores the importance of peripheral feedback in updating this position (Polit and Bizzi 1979) (see also Chapter 5).

The equilibrium point hypothesis as predicted by the mass–spring model therefore seems to hold, at least in limited experimental conditions. The Bizzi finding that deafferented animals can achieve the desired final position of a joint even when perturbations are applied is the strongest argument in favour of this hypothesis. Confirmation was obtained by Kelso and Holt (1980) in experiments dealing with human subjects. Subjects had to move one finger rapidly in order to reach a previously learned target position. A load could be unexpectedly applied to the moving finger. In both the normal and the perturbed conditions subjects were able to reach the target position with reasonable accuracy. In the same subjects, the moving finger was functionally deafferented by applying an inflated cuff at the level of the wrist. Movements executed under this condition, although they were globally less accurate than before application of the cuff, showed no difference in final position whether the load perturbation was applied or not. This experiment thus confirmed that proprioceptive information is not required for reaching a final position at a single joint and stressed the predominant role of motor programs in the generation of simple spatially coded movements.

Negative arguments relating to the equilibrium point hypothesis arose from several sources, First, some of the earlier results have proved not to be replicable. Day and Marsden (1982), for instance, in an experimental situation similar to that of Kelso and Holt (1980) trained subjects to flex the top joint of their anaesthetized thumb at a learned joint angle. These subjects were found to make large errors when a load was applied just prior to movement onset. Anaesthesia also resulted in reduction of the EMG response to the perturbation, the extent to which the EMG response was affected correlating well with the loss in accuracy. Day and Marsden's interpretation of these findings was that the EMG response to the perturbation normally provides effective compensation for the change in load, and therefore, that compensation reflects the activity of a closed-loop system rather than of an open-loop encoding of final position. Other more indirect refutations of the mass–spring model have been published. For instance, Larish *et al.* (1984) have shown that subjects failed to reproduce a learned position of their elbow joint by a rapid flexion movement when a 130 Hz vibration was applied to their biceps tendon prior to movement onset. Such a vibratory stimulation is known to alter position sense and to produce illusory displacements of the limb (Goodwin *et al.* 1972, and see Chapter 5). Larish *et al.* (1984) explained the subjects' failure by the distortion of afferent input documenting initial limb position, and concluded that accurate knowledge of initial conditions may be critical for limb positioning, in direct contradiction to the equilibrium point hypothesis. These contradictions led Bock and Eckmiller (1986) to propose that the controlled variable in programming movements should be movement amplitude rather than final position. Experimental data documenting their hypothesis are described in Chapter 3.

Finally, the main limitation of the equilibrium point hypothesis is that it holds only with a particular type of limb movement, that is, single joint, uniplanar movements. Most actions in fact imply movements involving several joints, they are executed in three dimensions, and have to deal with gravitational forces. It will be shown that in the situation of more natural and complex movements, the equilibrium point hypothesis fails to account for the observed motor patterns. In their experiments with single movements in deafferented monkeys, Bizzi and his colleagues had already noticed a number of facts that seriously questioned this mode of explanation and tended to show that, in addition to control of the final position, there were other active central processes controlling the course of the trajectory of the arm movement (for an account of these data, see Bizzi *et al.* 1982, 1984). The notion that the equilibrium point hypothesis cannot account for movement beyond a certain level of complexity and that programs must control more than a mere change in activation of the agonist and antagonist muscles, was in fact already suggested by the pattern of EMG discharges related to movements (the triple burst pattern). The burst on the antagonist muscle, as well as the second burst on the agonist, testify to the existence of additional central mechanisms for controlling the arrest of the movement and the postural state following the new position of the limb.

1.4. The programming of complex movements and action sequences

In this section the theoretical and empirical notions discussed for simple movements will be re-examined in relation to actions involving several musculoskeletal segments. In such complex actions, the motor commands issued to the moving segments are not accessible to the experimenter and have to be inferred from the study of the limb trajectories that they ultimately produce.

1.4.1. The contribution of N. Bernstein

Research in this domain has been largely influenced by the theoretical approach of N. Bernstein. Bernstein's main work was published around 1947 in Russian and did not become available in English until twenty years later (Bernstein 1967; for a new critical edition, see Whiting 1984; quotations given below are reproduced from the 1967 edition). Bernstein's premise was that 'the relationship between movements and the innervational impulses which evoke them is extremely complex and is, moreover, by no means univocal. A given impulse, he stated, may produce completely different effects under different conditions, because the response of the skeletomotor system to that impulse will depend on the initial position of the limb and on the external force field in which the movement develops. This conclusion was drawn by Bernstein from dynamic considerations showing that angular acceleration of a limb is determined by interplay between muscular forces acting upon it and its moment of inertia. In addition, the action of external forces (such as gravity) on the movement changes with the angle of the moving segment with respect to the fixed part of the limb. It follows that if a movement involves simultaneously several joints, computation of the forces acting upon it becomes so complicated that it cannot be foreseen by the central nervous system.

In addition, according to Bernstein, movement cannot be understood by the operation of a single impulse—it is the result of simultaneous cooperative 'systems of impulses'. The problem arises of whether these systems of impulses develop in time as a set of parallel sequences, or whether they are mutually interdependent. Bernstein thought it was possible to prove the homogeneity of a movement in the time domain, and that this particular homogeneity originated in the operation of the central nervous system. 'This demonstrates', he said, 'that there exist in the central nervous system exact formulae of movements or their engrams and that these formulae or engrams contain in some form of brain trace the whole process of the movement in its entire course in time.'(p. 37).

Bernstein suggested two possible modes of organization for the realization of motor sequences, or, in other words, two types of temporal structure for motor programs. The first possibility was that the engrams are chained, so that the stimulus for the occurrence of engram B is the existence of engram A. In that case, the temporal structure of a movement would be self-paced by the order of succession of the engrams that compose it. The other possibility implied a parallel release of the engrams, in the sense that there was no guiding principle for their succession. The latter structure was obviously more appealing to Bernstein. He postulated the existence of an additional engram which contained the entire scheme of the movement and guaranteed the order and the rhythm of realization of the scheme. An important characteristic of this 'motor image' was that it had to represent the form of the movement to be achieved, not the temporal sequence of neural impulses that produces it. This assumption, according to Bernstein, had the logical consequence that coding of the movement had to take into account proprioceptive input, where information about the external force field arises. It is therefore interesting to note that action-related proprioceptive input was central to Bernstein's hypothesis on motor programming. Proprioception, however, was not considered by Bernstein in the Sherringtonian sense of a reflex-triggering device, but rather in as much as it contributed to the central representation of the movement. This conception stands half-way between the centralist and the peripheralist theories of motor programming.

Finally, another important concern of Bernstein was the degrees of freedom problem. A solution to the complexity of peripheral organization of the motor system would be that movements are not completely determined by the central effector processes and that impulses are adapted by the peripheral afferents. In order to reduce the conceivably large amount of peripheral information to be processed, particularly in the case of complex movements, Bernstein assumed that many actions were organized as synergies (like walking, reaching, orienting, throwing), a type of organization which would reduce both the number of degrees of freedom to be controlled by the effector and the corresponding stream of afferent information.

1.4.2. Temporal coordination of commands in multisegmental movements

A decisive test for sorting out the two possible Bernstein models for generation of actions involving several limb segments, is to study the timing of commands sent to the corresponding muscle groups. There are a few examples of such studies, the outcome of which is much more in favour of relatively synchronous release of motor commands than of sequential activation. The first example is the timing of articulator movements during speech. It has been shown by cinefluorographic analysis that movements of the lips, tongue, velum, and jaw occur in highly synchronous patterns. Kent *et al.* (1974) have explained these data by postulating the existence of motor programs that would issue commands for several articulators simultaneously. These authors have used the notion of 'coordinated structure' to indicate that neural commands are temporally grouped, so that signals are sent to the muscles in a coherent fashion (for an account of the notion of co-ordinated structure, see below). This mode of programming may have the disadvantage of obscuring the phonetic boundaries (in contrast with other modes involving sequential activation for each phoneme), but it has the advantage of promoting articulatory efficiency. Similarly, the study of finger movements in skilled typists suggests considerable grouping of the motor commands during typing the single letters that compose a word (McLeod, personal communication, September, 1985).

Another example of synchrony of motor commands can be found in the action of catching a target by eye, head, and arm movements. Biguer *et al.* (1982) have measured the latencies of eye, head, and arm movements directed at the same target in normal subjects, and showed that the overt movements were arranged in a sequence where the ocular saccade was the leading event. The saccade was followed shortly by the onset of the head movement and finally by the onset of the arm movement. This pattern of a serial ordering of eye, head, and arm movements during orienting broke down when EMG activation, instead of the onset of movements was used as an index for latency. The latencies of the arm and neck muscle EMGs were found to remain within the same range as the latencies of the corresponding eye movements, so that activation of the three segments was in fact practically synchronous (for a complete description, see Chapter 2).

These examples stress the fact that the neural commands forwarded to muscle groups implicated in a particular action may be generated in parallel, even though the overt movements appear sequentially. Such a synchrony of neural commands for achieving a single final motor state (e.g. the articulation of a phoneme or reaching a target in space) is evocative of the notion of synergy held by Bernstein. Motor synergies have the already mentioned consequence of reducing the number of degrees of freedom within a given motor ensemble, by constraining groups of muscles to act within functionally coherent units. Beyond simplification, another consequence of grouping the commands is that the organization of the overt motor pattern is entirely determined by the 'passive' properties of the system (like inertial load of the different segments). In the case of the action of reaching a target, parallel and synchronous activation of eye, head, and arm musculoskeletal segments results in sequentially organized movements.

A more elaborate conception of motor synergies has been developed by Kelso and his colleagues. Kelso *et al.* (1979) recorded the movements of the two hands in subjects requested to point at two targets appearing simultaneously. They found that the beginnings and ends of the movements of the two hands tended to be synchronized. This fact remained true even if the two targets were of widely disparate difficulty (e.g. one large target located at a short distance from the home position for the right hand, and a small, distant target for the left hand). In that case, although the two hands moved at entirely different speeds to different points in space, times to peak velocity and acceleration for the two movements were relatively synchronous (Fig. 1.11). Such a synchrony of the two hands was quite remarkable since, when each movement was executed in isolation, large differences in duration (related to the distance of the target, as predicted from Fitts' law) became apparent. [Note, however, that significant differences in onset time and movement duration were found between the two hands in a similar experiment by Marteniuk *et al.* (1984)]. Kelso *et al.* (1983) observed the same pattern of coordination between the two hands in situations where an obstacle had been placed on the path of one limb, but not the other: in that case, according to the authors, space–time behaviour of the two limbs was affected and modulated in the same way and to the same extent.

Similar results have been obtained by Jeannerod (1984) in a different context—that of two-handed movements involved in manipulating a single object. Results indicated that arm movements for transporting the hands at the object location were initiated synchronously and reached their maximum velocity at the same time. In addition, formation of the two finger grips for grasping and manipulating different parts of the same object also had a common timing in spite of large differences in the sizes of the two grips (Fig. 1.12).

Observations like these not only suggest that anatomically independent musculoskeletal elements can become functionally (albeit temporarily) linked for the execution of a common task. The concept of 'coordinative structure' as used by Kelso *et al.* (1980b; see also Kelso and Tuller 1984) also implies that these temporarily constrained ensembles should behave dynamically in the same way as simpler functional units, namely, they should have the same tendency to self-equilibrium (Kugler *et al.* 1980). An application of this idea was developed by Arbib (1985). According to this author, multiple-joint motor patterns involve functional units (what he calls 'virtual' segments), different from the elemental musculoskeletal segments but which, when activated, behave temporarily like single segments. In prehension tasks for example, fingers group together according to the shape or the weight of the target object. The number of fingers that group together (the size of the 'virtual finger') seems to relate exclusively to the amount of opposing force required to raise the object. This 'co-ordinated control program' approach has proved to be useful in designing computer programs for controlling robots (Arbib 1985).

The notion of co-ordinative structure has important implications for the programming of complex movements. First, because it predicts that movement is specific to the unique requirements that arise from interaction between the subject and the goal, it does not imply detailed programming of aspects of the movement like trajectory shaping, for instance. As Arbib (1981) argues, motor representation can be conceived as consisting of 'units of knowledge' of interactions between the subject and the environment (what he calls perceptual and motor 'schemas'). Motor learning could result from increase in knowledge about the environment or about the conditions in which movement is performed. The second implication of the

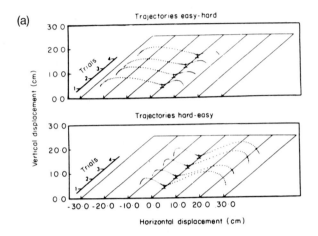

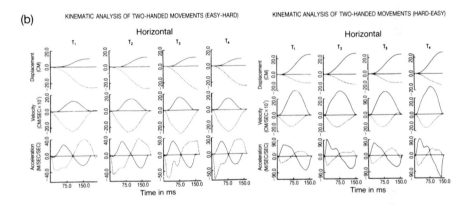

Fig. 1.11. Pattern of co-ordination in two-handed movements.

(a) Movement trajectories have been recorded two-dimensionally in subjects pointing with the two hands at two different targets. The two targets are located at the same or different distances from starting position (at midline). Pointing at a target close to the midline is considered an 'easy' task. Pointing at a target far from the midline is considered a 'hard' task. Movement trajectories have been recorded in conditions of different degrees of difficulty (easy–hard, hard–easy) for the two hands.

(b) Kinematic analysis of four trials (T1–T4) involving 'hard' movements with the right hand (solid lines) and 'easy' movements with the left hand (dotted lines). Note relative synchrony of the two hands, more clearly seen on the acceleration profiles (lower row). (From Kelso *et al.* 1983, with permission.)

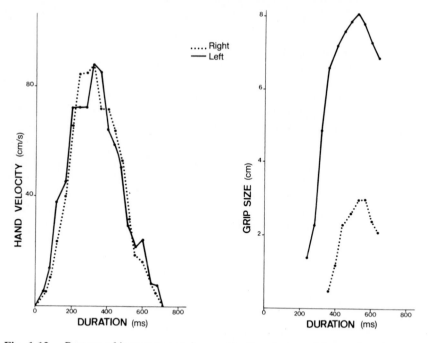

Fig. 1.12.   Pattern of intersegmental co-ordination during a bimanual prehension movement. The subject had to reach for a small bottle placed 40 cm in front of him and remove the plug placed at its top. This action required transporting the two hands at the same location and performing two different grasps. In this case, the right hand (dotted lines) grasped the plug (0.6 cm) and the left hand (solid lines) grasped the bottle (6.5 cm).

Left: velocity profiles of hand displacements from resting to target positions. Note synchronous onset, peak velocity and termination of the movements of the two hands. Right: size of the finger grips for the two hands. Note synchronous changes in size, in spite of widely different finger apertures. The initial parts of the finger grips could not be plotted due to shadowing on the film. For details on prehension movements, see Chapter 2. (From Jeannerod 1984.)

co-ordinative structure concept is that, because it predicts that the motor ensemble tends to self-equilibrium, it explains rapid compensation for perturbations applied to any part of the system (Saltzman and Kelso in press; Kelso *et al.* 1984).

Movements of articulators during speech provide good examples that illustrate this conception. Perturbations applied to one articulator during production of a given phoneme are apparently immediately compensated by the other articulators, in such a way that production of the target phoneme is unaltered by the perturbation. As shown by Folkins and Abbs (1975), perturbations applied to the jaw during jaw elevation movements are compensated by lip muscles, so that lip closure can be achieved normally. Similarly, perturbations to the lower lip can be compensated by the upper lip (Abbs and Gracco 1981). These compensations which, again, represent evidence for motor equivalence, do not seem to be based on a simple feedback system, but rather seem to act via an 'afferent-dependent feedforward process' (Abbs 1982) (this point is developed further in Chapter 5). Response latencies for compensations in the oro-facial musculature range from 22 to 75 ms, i.e. they are different from responses pertaining to brain-stem reflexes (around 15 ms) (see also Kelso *et al.* 1984). Abbs and Gracco (1984) concluded from these observations that the compensatory mechanism is more likely to be an open-loop predictive mechanism, than a closed-loop corrective one.

It is interesting to note that in both the co-ordinative structure and the co-ordinated control program theories, the role of sensory feedback from movement execution is conceived in a very similar way as in the Bernstein model. The role of feedback is conceived here as that of an optimizing device, in the absence of which the whole action can still be executed (though with imperfect timing), and not as a trigger for producing motor sequences, in contradistinction to earlier models.

1.4.3. Temporal regularities in complex actions

Relative stability in the time domain seems to be another basic principle for programming movements involving simultaneously several joints, a principle which has already been discussed for single-joint movements earlier in this chapter. In the case of complex movements, such a 'proportional duration model' (Gentner 1985; see Schmidt 1982) predicts that the durations of all the components of a motor sequence should maintain a constant proportion of the overall duration, as the overall duration of the sequence changes.

There are situations where this prediction is fulfilled and where the time for executing a given task is maintained approximately constant on each repetition, with the corollary that velocity of each component is rescaled according to the demand of the task. Indications of this type of temporal regularity in motor behaviour have existed in the literature for almost a hundred years (Binet and Courtier 1893), yet its importance has been recently re-emphasized. Studies of writing movements, for instance, show that it takes the same time to write a letter or a word at different sizes, which implies proportional changes in velocity (Michel 1971). This relationship between extent of movement and velocity in writing (the 'isochrony principle' of Viviani and McCollum 1983) appears to be a rather common feature, not only in writing but also in actions like typing (Viviani and Terzuolo 1980), lifting a weight (Gachoud *et al.* 1983), or grasping objects (Jeannerod 1984). Duration invariance has important implications in determining the intrinsic kinematic structure of the movements. In the action of writing, the fact that movement duration tends to remain constant when extent varies results in constancy of the shape of the velocity profile for a given word written at different sizes (Fig. 1.13). The same is observed if a given word is written at the same size but at different speeds: although in that case total duration varies, temporal landmarks on the velocity profile (e.g. velocity peaks) remain scaled to each other, in fulfilment to the proportional duration model. Similarly, in typing, relative timing of the keystrokes was apparently maintained as the overall duration of the word changed (Viviani and Terzuolo 1980). This mode of programming has the consequence of keeping the timing of the commands independent of the spatial parameters of the movement. In other words, the selection of the muscles to be activated for achieving a given task can be modified (according to the motor equivalence concept), or the torque applied to the joints can be modulated, without affecting the temporal template that determines the co-ordination for a given action.

Gentner (1985), however, in re-examining data from the literature, concluded that there was less evidence for a proportional duration model than commonly accepted. One of his reasons for rejecting the model was the large temporal variability observed in performing the same movement across different trials. In a series of typing experiments, Gentner (1985) found that the durations of interstroke intervals during several types of the same word were not correlated to each other, and only loosely correlated to the overall word duration. These results are in apparent conflict with those of Viviani and Terzuolo (1980), although it should be noted that Viviani and Terzuolo did not analyse their data statistically. It follows that, as for simple movements, the hypothesis of a generalized motor program for action sequences, which would require a proportional duration of the components of the sequence, is not clearly supported by the experimental data.

1.4.4. Trajectory formation

Natural movements involving several joints are characterized by the trajectory of the end point of the limb. In the case of arm movements, trajectory refers to 'the path taken by the hand as it moves to a new position and the speed of the hand as it moves along the path' (Abend *et al.* 1982, p. 331).

The problem arises of what aspect of the movement is actually represented in the program in order to produce the required trajectory. There are several alternatives. On one hand the equilibrium point hypothesis predicts that only the final hand position has to be planned and that the trajectory of the hand will be 'passively' determined by inertial and viscoelastic properties of the musculoskeletal system of the arm. On the other hand, the phase relationships of the displacements of the joints participating in the movement, and ultimately producing the trajectory, might also be

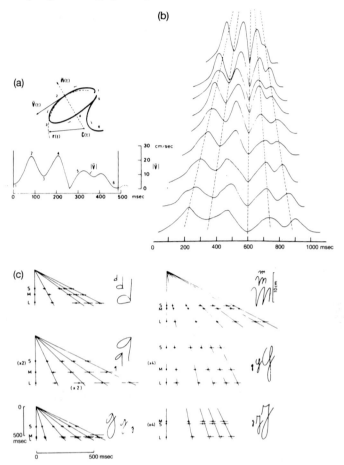

Fig. 1.13. Spatio-temporal regularities in handwriting.

(a) Dynamic description of handwriting. The change in tangential velocity of the stylus has been plotted against time (lower diagram) during writing the letter 'a'. Features of the velocity profile, identified by corresponding numbers on the trajectory, can be used to define a pattern for the movement.

(b) The same subject as in (a) was required to write letter 'a' at increasing writing speeds while keeping constant the size of the letter. When the total duration of the movement decreases, the instantaneous values of the velocity increase proportionally in such a way as to leave invariant the ratios among the times of occurrence of the major features of the profile.

(c) Letters were written at three different sizes—small (S), medium (M), and large (L)—by two subjects (left and right columns). For each action of writing a letter, temporal features were selected on the tangential velocity profiles and plotted as a function of time. It was found that, in spite of different durations, a relationship was maintained between temporal features across different sizes, as shown by the straight lines connecting the same features. (From Viviani and Terzuolo 1980, with permission.)

represented. Study of single-joint movements does not allow one to disentangle the two alternatives, for the obvious reason that only one trajectory between the resting and the final positions is possible. By contrast, in multiple joint movements, the relation between trajectory of end-point of the limb and the angles of the joints is not univocal. This problem was studied by Morasso (1981) in an experiment where subjects had to displace a lever in the horizontal plane between visual targets. The movement involved flexion–extension at the elbow and shoulder joints. Results showed that the hand was displaced from target to target along relatively straight lines and that for a given subject movements had a fairly constant duration whatever their direction or extent. The curve of the tangential hand velocity versus time had a bell-shaped pattern with a single peak of velocity and this shape was maintained across movements of different directions and extents. By contrast, the joint angular velocities differed widely according to the direction of the movements.

The fact that target-oriented movements have straight trajectories (although the same movements executed in the absence of a target have curved trajectories) now seems well established. Abend et al. (1982) have shown that when subjects are required specifically to produce curved trajectories they tend to segment their movement in quasi-rectilinear subunits. In such cases, the tangential velocity profile of the hand presents as many velocity peaks as there are straight segments, that is, the velocity decreases at each change in direction of the trajectory (see also Flash and Hogan 1985). A very similar finding was reported by Viviani and Terzuolo (1980, 1982) in their study of hand trajectories during writing. These authors have shown that in this complex type of movement, the tangential hand velocity was an inverse function of the degree of curvature, such that velocity increased during straight segments and was at its minimum in the curved parts. It should be mentioned, however, that the notion of straight trajectories does not account for movements performed in three dimensions. In that case, movement trajectories are usually curved in the vertical (gravitational) dimension without apparent segmentation in the velocity profile of the movement (Jeannerod 1984).

The problem thus remains to decide which aspect, the final position of the end-point of the limb or joint angles, is the most relevant for describing the program of complex movements. Morasso (1981) interpreted his own findings as indicating that movements are programmed in terms of kinematics of the end-point of the limb in space, not in terms of joint angles or muscle addresses. Thus, determining the position of the end-point of the limb in extrapersonal space might be the critical factor for trajectory formation. In confirmation of this hypothesis, Soechting and Lacquaniti (1981) have shown that, in pointing movements involving forward flexion of the shoulder and extension of the elbow and executed at different speeds, the trajectory of the end-point of the limb in space varied little from trial to trial, and remained independent of the speed of the movement. Since executing the same movement at different speeds involves large changes in torques applied to the joints, invariance of the trajectory implies that it is this aspect of the movement which is planned and controlled. As emphasized by Soechting and Lacquaniti (1981),

a movement at a given speed cannot be produced simply by an appropriate scaling in amplitude and time of the requisite torques acting at the shoulder and elbow. The net torque acting at each joint is a combination of gravitational torques (related to angular displacement and independent of speed), Coriolis forces (proportional to the square of angular velocity), and inertial torques (proportional to angular acceleration). (1981, p. 718)

In another experiment Lacquaniti and Soechting (1982) have studied reaching movements involving simultaneously the same coordination between elbow and shoulder as in the previous experiment, and rotation of the wrist. They found that the relationship between the shoulder and elbow angular motions (and therefore the trajectory of the end-point of the limb) remained invariant over a wide range of movement speeds. By contrast, no such relationship existed between angular motion of the wrist and the motions at the other two joints. This finding seems consistent with the fact that in this particular task, the wrist motion did not influence the position of the end-point of the limb and was not involved in trajectory formation: consequently it was likely to have been planned separately (Fig. 1.14).

The fact that it is the movement itself and not the pattern of activity in individual muscles that is invariant in multisegmental actions like reaching seems to be demonstrated by a third experiment by the same group (Lacquaniti *et al.* 1982). In this experiment shoulder–elbow pointing movements were executed under conditions of load (subject carried a 2.5 kg weight in his hand), or change of arm length (by attaching a rod to the arm). Results showed immediate compensation for these perturbations, i.e. arm trajectory remained unchanged in spite of large changes in muscle activity.

These results raise the interesting problem of the system of co-ordinates by which a goal-directed movement is organized. Soechting and Lacquaniti (1981) rightly made the remark that programming a movement in terms of 'intrinsic' co-ordinates (joint angles, relative joint angular displacements) would imply storage of a large number of combinations between the joints involved. The above results suggest a more economical solution, whereby trajectories would be planned according to a mapping of the target position in space in an 'extrinsic' co-ordinate system. This hypothesis is close to that of spatial control postulated by Morasso (1981).

These considerations, however, do not solve the problem of what parameters are actually represented at the program level. Target position is an abstract concept in the motor domain. In order for target position to be transferred into an accurate movement, the spatial 'map' where target position is encoded has to select and activate the appropriate muscles for

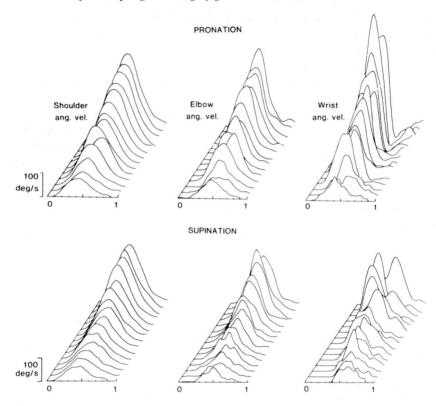

Fig. 1.14. Co-ordination between joints during reaching. Shoulder, elbow, and forearm angular velocity profiles in one subject. In the upper row, single trials involving pronation are plotted in ascending order of shoulder velocity; in the lower row, trials involving supination are shown. All trials have been normalized in time relative to the duration of shoulder flexion. Note the shorter duration and the greater variability of wrist motions compared with the other two motions. (From Lacquaniti and Soechting 1982, with permission.)

giving the movement its correct direction. Simultaneously, a distinct mechanism must determine the distance to be covered along the selected path, by controlling the appropriate movement parameters (e.g. duration and velocity) in the time domain. Separation of the mechanisms for direction and amplitude of movements was postulated by Bernstein in 1967 (the ecphorator concept). Other more recent models have also maintained that these two mechanisms are separated (see Arbib 1981; see also Chapter 2).

1.5. Conclusion

In this chapter some of the models used for explaining the generation of goal-directed movements have been reviewed. The concept of the motor program stands at the junction between several levels of explanation. On the one hand, it can be conceived as a detailed representation of movement parameters, the unfolding of which will produce the motor pattern. This approach has taken advantage of neurophysiological descriptions of central and peripheral events related to preparation and execution of movements. The existence of clear-cut and rather stereotyped patterns of EMG discharges for each individual movement has encouraged the search for simplified models where parametrization of the movement would be reduced to a few variables such as force or duration of the neural activation corresponding to these discharges. However, the expectation that the engram would represent a predictable force–time function applicable to all movements of a given class, has proved to be largely illusory. Movements are not a transparent phenomenon, in contrast to what theories stressing a direct perception-into-action transformation would imply.

Reduction of the number of controlled variables remains a general trend among motor theorists. Models limiting the role of the program to a readout of the spatial map where target position is encoded, also fulfil, at least in theory, the same criterion of economy. These models postulate that both trajectory and kinematics of a movement can be specified by the passive response of the musculoskeletal system to a change in muscle tension corresponding to the spatial location of the goal. They would have difficulties, however, in explaining such common findings as cognitive penetrability and contextual coding of movements. In addition, they do not appear to hold in situations of complex movements.

Programs for complex movements can thus hardly be conceived in terms of a detailed programming hypothesis. According to such a hypothesis, the finding that even complex movements have straight trajectories would imply a detailed coding of the respective joint angles as a function of time, which might prove extremely costly in computational terms. By contrast, programming limited to the final position of the end-point of the limb would produce a resultant trajectory representing an optimized displacement of the involved segment, where the apparent co-ordination of the joint angles would be incidental. The same logic applies to the sequencing of segmental movements in complex actions like orienting, reaching, or speaking: the apparently subtle timing between segments in fact reflects a much simpler organization principle of grouping and synchronizing the commands. The notion of momentary and flexible structures constraining the segments involved in a given movement to act together, which has been applied to account for these facts, seems an interesting one. It implies that the co-ordinated program that rules the structure generates higher-order commands, such that common kinematics and perhaps a common tendency to equilibrium would ensue for all the segments which comprise the structure.

These conceptions of motor programming have in common an emphasis on the *goal of the action* (of which the final position is merely a kinematic metaphor) as the most likely element to be represented at the central level. Execution of the movements could then be ruled by economy principles and organized so as to reduce energetic cost. Mechanical constraints of motor systems could be taken advantage of for conserving part of the energy, by trading it between its kinetic and potential forms. These principles which seem to account for execution of periodic motor activities (such as locomotion, e.g. Taylor *et al.* 1980), might also prove operational for discrete movements. This line of research is still largely open to theoretical and empirical validation.

2. The hierarchical organization of visuomotor co-ordination

The emphasis on movement accuracy, which has represented the bulk of behavioural studies of movement until recently, has favoured the trajectory of the end-point of the limb as the main aspect to be controlled by central and peripheral motor mechanisms. This conception is reflected by the extensive use of single-joint movements in the study of motor control, and also by the tendency to reduce complex movements to units functioning in the same mode as single joint movements. Although this view may be justified in theory (see Chapter 1), it has resulted in the neglect of the study of co-ordination between body segments. Parameters currently considered in the study of single-joint movements, such as latency, duration, error with respect to the target, and kinematics of the end-point of the limb, are far from representing an adequate description of motor performance, especially for complex, multisegmental actions.

In the present Chapter, an essential aspect of movement control will be considered, namely the spatial and temporal organization of actions involving several segments of the musculature. In the context of the action of reaching, the co-ordination of eye, head, and arm movements during orientation, and the co-ordination of proximal and distal limb segments during prehension are the main motor ensembles to be considered.

2.1. The co-ordination of eye, head, and hand movements during reaching for a visual target

2.1.1. Eye–head co-ordination

The respective timing of eye and head movements during acquisition of a visual target is such that in the overt sequence of movements, the eye movement usually has the shortest latency and the eye precedes the head (Bartz 1966). In addition, the eyes, which move with greater velocity, reach the target first and foveate it before the head has finished moving. As a consequence, the orienting movement of the eyes has to be followed by a movement of the eyes with respect to the head, to maintain gaze fixation on the target during continuing head rotation. Although this sequence seems to be the most frequently observed in experimental conditions where a target appears suddenly in the peripheral visual field, other eye–head strategies have been described. For instance, in situations where the spatial position of the target is predictable, the head movement latency

41

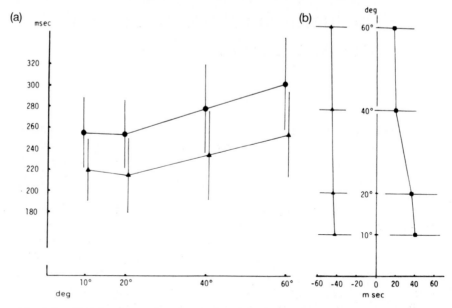

Fig. 2.1. Eye–head co-ordination during orienting movements in humans.

(a) Mean saccadic eye movement latencies (triangles) and neck EMG latencies (circles) have been measured in ms as a function of target distance (in degrees). Note consistent eye–neck EMG delay of about 40 ms for the full range of target distances.

(b) Relation of neck EMG latencies and head movement latencies to saccadic latencies during orienting movements. The vertical line perpendicular to the time axis represents saccade onset (0 ms). Neck EMG activity begins about 40 ms prior to saccade onset. Head movement onset is delayed by about 20–40 ms with respect to the saccade. Note neck EMG–head movement delay longer in movements directed at closer targets (10°, 20°; see Fig. 2.3). (From Warabi 1977, with permission.)

may be shorter than that of the eye (e.g. Bizzi *et al.* 1971, 1972 a Zangemeister and Stark 1982).

In the 'triggered' mode of eye-head coordination, (i.e. the ocular saccade precedes the head movement), activation of the neck muscles occurs some 20–40 ms prior to the eye movement, whatever the eccentricity of the target with respect to the initial fixation position (Warabi 1977; Zangemeister and Stark 1982) (Fig. 2.1). In the study of Biguer *et al.* (1982, 1985), the latency of the neck EMG also was found to be slightly shorter than eye movement latency. The difference in latency reported by the latter authors, however, appeared to be statistically significant only for targets located 40° from the midline, and not for targets closer to the midline. In addition, for targets located at 10° from the midline, the pattern reversed and eye movement latencies were found to be systematically shorter than

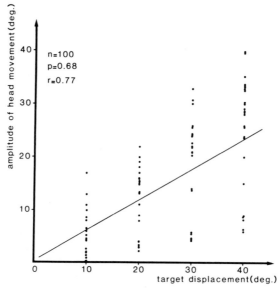

Fig. 2.2. Hypometry of head movements during orientation at targets within the peripheral visual field. The diagram shows the relationship between amplitude of head movement (in degrees of arc) and target displacement from the midline. Each point of the scatter represents an individual trial in the head-free condition (n = total number of individual trials; p = slope of regression line; r = correlation coefficient). (Data from 5 subjects, from Biguer *et al.* 1985.)

those of neck EMG (see below, Fig. 2.3). Taken together, these results in humans confirmed previous observations in monkeys, showing that EMG activity of the ocular muscles consistently preceded EMG activity of the neck muscles by about 20 ms as shown by Bizzi *et al.* (1971). In the same study, Bizzi *et al.* showed that all neck muscles involved in head turning were activated simultaneously irrespective of the initial head position. This synchrony may reflect the fact that neck muscles are not only involved in head turning, but also have an important function in the maintenance of posture.

Another interesting aspect of eye–head co-ordination is the relative eye and head positions that are achieved at the end of the eye–head sequence. The amplitude of the head movement is usually hypometric with respect to the target position. Gresty (1974) found that although final head positions were approximately linearly related to target positions, the head movement accounted only for about 75 per cent of the target amplitude. This finding was replicated by Biguer *et al.* (1984, 1985), who found that in target-oriented movements the mean final head position was about 5° for targets located at 10° from the midline, and about 12°, 19°, and 25° for targets located at 20°, 30°, 40° respectively, from the midline (see also Herman *et al.* 1981). The amplitude of head movement nevertheless correlated positively with target position (Fig. 2.2). These results mean that the head axis is almost never aligned with target position, and that a large amount of the distance to the target (about 40 per cent in the Biguer *et al.* experiment) has to be covered by a displacement of the eyes with respect to the head. The same pattern was also observed in monkeys (Bizzi *et al.* 1971; Tomlinson and Bahra 1986). In the cat, however, the head seems to be constantly aligned with the target at the end of the orientation movement (see Chapter 4).

It is interesting to mention that the eye–head behaviour can be modified with training. According to Roll *et al.* (1986) certain subjects trained for high accuracy in sport actions like hitting a target or throwing a ball move their head more than untrained subjects. When they are tested in an experimental situation involving accurate aiming at targets within the peripheral visual field, they make combined eye–head movements for all target excentricities, including small ones, and the amplitude of their head movements covers most of the distance to the targets.

Hypometry of head movements during orienting suggests that eye–head coordination is not a unitary mechanism, but rather the summation of several mechanisms. One of those would relate to fixating objects within the near periphery of the visual field, and would require primarily eye movements with only little contribution from the head. Another mechanism would relate to bringing the eyes to or near more remote targets, and would require moving both the eyes and the head in a co-ordinated way. In fact, even a third mechanism could be thought of, when eccentricity of the target is such that not only eyes and head, but also the trunk rotates. Arguments for these possibly separate mechanisms can be drawn from animal experiments with stimulation of the superior colliculus. In the cat, stimulation of the rostral part of the colliculus produces eye movements alone, the amplitudes of which do not exceed 25° from the central position. By contrast, stimulation of the caudal part produces systematic eye and head co-ordinated movements (Harris 1980; Roucoux *et al.* 1981).

The possible duality of mechanisms for orienting has important implications concerning the role of eye and head position signals in specifying the direction of reaching movements. This point is fully discussed in Chapter 4.

2.1.2. Eye–hand co-ordination

In the action of reaching to a target by hand, the overt hand movement has been shown to lag the eye movement. According to different authors, the mean eye–hand delays range from 60 ms (Angel *et al.* 1970) up to more than 100 ms (Prablanc *et al.* 1979a; Herman *et al.* 1981). Differences in values between authors can be explained by technical or procedural differences. Angel *et al.* used the arm acceleration signal as an index for arm latency, although Prablanc *et al.* used the onset of arm movement [see Megaw and Armstrong (1973) for an account of this difference]. The experimental condition in which the eye–hand delay is measured also seems to be an important factor in creating differences in the values of the eye–hand delays. Prablanc *et al.* (1979a) showed that the amount of visual information provided to the subject during the movement had an influence on both eye and arm latencies. As an example, in reaching toward a target located 20° from the midline, the fact of shifting from a condition with visual feedback to a condition without visual feedback produced an increase in arm latency by about 30 ms (from 348.3 ms to 375.5 ms). The eye–hand delay increased from 114 ms to 127 ms, indicating that the eye latency was less affected than the arm latency by the difference in experimental conditions. This result was confirmed by Herman *et al.* (1981).

If EMG activation of arm muscles, instead of arm acceleration or displacement, is used as an index for arm latency, a new pattern of eye–hand coordination emerges. Biguer *et al.* (1982, 1985) showed that, in pointing movements, the EMG activation of arm muscles (e.g. the biceps brachialis, or the deltoid) was synchronous with the onset of the corresponding ocular saccades. This result (see below) suggests that the commands for the eye and arm movements are released synchronously and that the 60–100 ms delay measured between the two represents the time needed for coupling biomechanically the muscular contraction and the corresponding limb displacement. Angel (1975) has found delays between EMG activation in arm muscles and arm displacement that are compatible with this hypothesis.

There are other indications for eye–hand coupling in target-oriented movements. For instance, in the Prablanc *et al.* (1979a) experiment, where the right arm was used for pointing at targets, arm latencies were found to be longer for movements directed at targets appearing within the contralateral (left) hemifield. The same effect was observed for the latencies of the corresponding eye movements, which were also longer (by about 25 ms) for targets located within the left hemifield. This result has recently been confirmed and expanded by Fisk and Goodale (1985). They found that hand movement latencies during pointing at visual targets were shorter when the target was located in the visual field ipsilateral to the hand. This held true for both right and left hands. However, overall right hand latencies (for movements directed at any target) were systematically shorter than left hand latencies. Eye movement latencies followed the same trend, that is, they were shorter when the saccades were directed at targets ipsilateral to the arm (right or left) used for pointing; and they were shorter when the pointing movement was made with the right arm.

These findings raise an interesting point. The fact that, according to Fisk and Goodale, saccades directed at the same target (e.g. 10° to the right of midline) have different latencies depending on whether they are coupled with a movement of the right (ipsilateral) or the left (contralateral) hand, strongly suggests that the asymmetry in saccade latency is a consequence of the coupling of the two movements. Asymmetry of eye movement latencies has been mentioned (in a context different from that of manual pointing at targets) by several authors. Pirozzolo and Rayner (1980) and Hutton and Palet (1986) found that visually oriented saccades in right handers had shorter latencies when they were directed from left to right than from right to left. An inverse effect was found in left handers. This asymmetry, however, cannot account for the results of Prablanc *et al.* (1979a) and Fisk and Goodale (1985). As all the subjects in the latter two studies were said to be right handers, they should have had shorter eye movement latencies for targets appearing to the right, irrespective of the hand used for pointing. Since this was not the case, the most likely explanation at this point for the observed latency asymmetry remains the coupling with hand movements.

Eye–hand coupling should normally imply a certain degree of correlation (on a trial by trial basis) between latencies of eye and arm movements. Prablanc *et al.* (1979a) found that the two latencies were only weakly correlated (correlation coefficients were around 0.5). A similar result was reported by Biguer *et al.* (1982), who used the biceps EMG as an index for arm latency, and by Gielen *et al.* (1984) who reported a correlation coefficient of 0.57. These findings can be interpreted as supporting the hypothesis of a parallel activation of the eye and arm motor systems (for a discussion of this point, see below). Other authors, however, have reported a much stronger correlation between latencies of the two movements ($r = 0.8$, Herman *et al.* 1981 in humans; $r = 0.9$, Rogal *et al.* 1985 in monkeys). The difference with the above-mentioned studies may be explained by the degree of training of the experimental subjects. Although Prablanc *et al.*'s subjects were entirely naïve, Herman *et al.*'s subjects were given a training session to stabilize their reaction times. Similarly, it is likely that the monkeys in the Rogal *et al.* experiment were highly trained, since they performed several thousand trials. In fact, another study by the Fischer group indicates that, under certain conditions, eye and head latencies can be clearly decoupled. In this study, subjects were instructed to make target-oriented saccades with very short latencies (*c.* 120 ms, 'express saccades'; Fischer and Ramsperger 1984). If hand pointing movements were made in the same trials, the arm latency did not get shorter than in the normal condition and the co-ordination between eye and hand latencies became weaker (Fischer and Rogal 1986).

2.1.3. Eye–head–hand co-ordination

From the above-mentioned data, one gains the impression that commands forwarded to the different muscular groups involved in the action of pointing are clustered within a relatively short span of time. Definite conclusions on this timing, however, can only be drawn from a recording of the complete set of events in the same subjects and during the same sessions. An experiment along these lines, already briefly mentioned in Chapter 1, has been carried out by Biguer *et al.* (1982, 1985). Subjects were required to point at targets using the eye, head, and hand as quickly and accurately as possible. In addition to eye movements and hand positions, head movements were recorded by means of a low-weight helmet secured to the subject's head and connected to a potentiometer, and by a sensitive accelerometer fixed to the helmet. Finally, EMG activity of the right posterior neck muscles, and the right arm (biceps brachialis) or shoulder (deltoid) muscles was recorded. Time was measured between onset of each target and the following events: ocular saccade, head movement, hand movement, biceps or deltoid EMG, and neck EMG.

In all subjects, the ocular saccade was found to lead the sequence. Its latency tended to increase with distance of the target from the midline, in agreement with previous findings (e.g. Bartz 1966; Prablanc and Jeannerod 1974). By contrast with eye movements, the latency of head movements tended to decrease with target eccentricity. This fact resulted in eye–head delays being longer when the pointing movement was directed at a target close to the midline, than when it was directed at a remote target (see also Uemura *et al.* 1980). Finally, the onset of hand movements had a constant latency of about 350 ms over the full range of target distances (Fig. 2.3(a)). Note that the latter result is in discordance with those reported by other authors showing that hand movement latencies also increase with eccentricity of the targets (Prablanc *et al.* 1979a; Herman *et al.* 1981).

This pattern of a serial ordering of eye, head, and arm movements during pointing broke down when EMG activation, rather than onset of movements was used as an index for latency. The latencies of biceps (or deltoid) and neck EMG were found to remain within 200–220 ms, and to be uninfluenced by target distance from the midline. Moreover, these EMG latencies were within the same range as the latencies of the corresponding eye movements. They were slightly shorter than saccadic latencies for targets located at 20° from the midline and beyond, although the reverse was observed for targets located at 10° (Fig. 2.3(b and c)). Finally, it was noted that the availability of visual feedback from the moving hand had no influence on this temporal pattern.

The above results have two main implications for the mechanism of eye–head–hand co-ordination. First, they suggest that inertia is the important factor in producing temporal sequencing of latencies for eye, head, and arm movements directed at the same target. Zangemeister *et al.* (1981) argued that differences in inertia might also account for the differences in the dynamic main sequence profiles of these three movements (Fig. 2.4). Second, these results tend to confirm the idea that neural commands forwarded to muscle groups implicated in the same action of pointing are generated simultaneously. This pointing 'synergy' would be under the dependence of a common generator, containing information as to location of the target in the body-centered space. The requirements for a common generator would be met by the eye movement generator. Eye movements

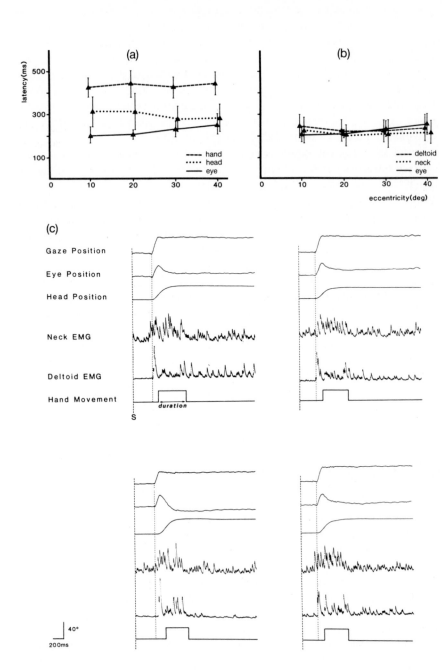

are directly coded in spatial co-ordinates, which implies that target co-ordinates on the retina are transformed into head-centred co-ordinates in order for the eye to reach the target location in space whatever the head position (see Chapter 4). It could therefore be argued that the signal used for this transformation could be available also for directing the head and arm toward the target. The hypothesis of the eye movement generator as a common generator, however, is not completely confirmed by the available data. First, EMG latencies for arm and neck muscles can clearly be shorter than eye movement latencies, even after subtraction of the delay for isometric contraction of extraocular muscles (This delay is of 7 ms in the monkey; Fuchs and Luschei 1970; Robinson 1970). Second, if commands were produced by a common generator, one should expect a high degree of correlation between EMG latencies and eye movement latencies. As mentioned earlier, this correlation was found to be rather weak, at least in normal conditions.

Rejection of the common generator hypothesis should logically yield to that of multiple, parallel spatial coders for eye, head, and arm positions. At present, however, there are no further arguments for proving or disproving either one of these hypotheses.

The same results have another interesting aspect. The relative synchrony of neural commands for eye, head, and hand muscles might have the function of producing a sequence of the overt movements. This sequence, in allowing time for correction signals to be used, might in turn have important implications for accuracy of the hand movement. The role of visual feedback has been shown to be critical for achieving terminal accuracy (see Chapter 3), and the fact that the eyes already foveate the target when the hand is approaching it creates a very favourable situation for using a fine-grain visual feedback signal.

Fig. 2.3. Eye–head–hand co-ordination during orienting movements.

(a) Mean latency (in ms) of eye, head and hand movements as a function of target eccentricity (in degrees of arc). Note longer head latencies for short target distances, a result similar to that of Warabi (1977) (see Fig. 2.1). Because in the present experiment head movements were detected with a sensitive accelerometer, this effect of target distance on head movement latency must be attributed to intrinsic factors rather than to detection problems.

(b) Mean latency of eye movements, right neck muscles EMG and right deltoid EMG during the same orienting movements as in (a).

(c) Four individual trials of orienting at a 40° target in the same subject. Eye position in the horizontal plane was recorded by EOG. Head position in the horizontal plane was recorded through a potentiometer affixed to the head. Gaze position was reconstructed electronically from eye and head positions. Beginning and end of hand movement were obtained from contact of the hand with the working plane. EMGs have been rectified and integrated with a short time constant. The vertical dotted line indicates onset of eye movement. Note relative synchrony of EMGs with eye movement.

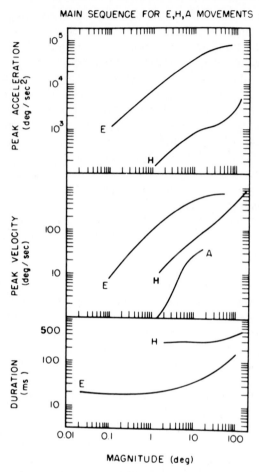

Fig. 2.4.   Comparison of main sequence data for eye, head, and wrist movements. Acceleration, velocity, and duration of movements have been plotted as a function of amplitude. Note important scaling differences in the three different biomechanical systems, as well as similar dependence of all three systems on movement amplitude (E = eye, H = head, A = arm). (From Zangemeister *et al.* 1981 with permission.)

2.1.4. Implications of eye–head–hand co-ordination for visuomotor behaviour

It has been suggested that gaze orienting is not only the leading event in the sequence of reaching (as described in the previous paragraph), but might also represent a precondition for execution of correct reaching. Arguments to support this hypothesis can be drawn from studies of visuomotor development and from studies of plasticity of visuomotor co-ordination in adults.

(a) Visuomotor development

Many studies in humans as well as in animals indicate that acquisition of accurate reaching depends upon previous acquisition of a stable target-oriented eye–head co-ordination. Human infants do not start to develop reaching before the age of about 20 weeks, that is, until neck muscles acquire sufficient strength. Not surprisingly, it has been observed that visuomotor behaviour in babies at the pre-reaching stage (5–8 weeks of age) can be greatly improved if their head is held and stabilized by an experimenter (Grenier 1981). Fine, co-ordinated and accurate goal-directed movements can be observed in this condition (Grenier's 'motricité libérée'), which contrast with the naturally awkward and somewhat erratic movements of babies at this age. This observation emphasizes the role of head position in building an efficient reference system for directing movements toward extrapersonal space (this point will be developed at length in Chapter 4).

In animals, systematic experiments by Hein and his colleagues have led to the notion that, during development, visuomotor co-ordination builds up in an orderly fashion. By using simple visuomotor tests like visually elicited and visually guided placing reactions, these investigators established the necessary conditions for a kitten to acquire visually guided reaching. Normal development requires not only that the kitten can see its forelimbs, but also that it can use visual feedback from its forelimbs in the context of locomoting. If the kitten is exposed in a visual environment during locomotion without view of its forelimbs (for example, if the forelimbs are hidden by an opaque collar) it will not develop normal visually guided reaching with either limb (Hein and Diamond 1972). This experiment demonstrates that there are constraints on the sequence in which visually guided reaching can be acquired. More specifically, it suggests that axial and proximal visuomotor components of reaching must develop first in order for the more distal components to finally take place. This point was clearly confirmed in another experiment where kittens were first exposed to a normal visual environment with one eye surgically immobilized (the other eye was closed). Formal testing of visually guided behaviour was made after two weeks using an obstacle course and a test of visually guided placing of one forelimb on a narrow bridge. These animals failed to show visually guided locomotion or visually guided reaching, and apparently were unable to develop this capacity until the experiment was terminated one year later. This result (Hein *et al.* 1979; Hein and Diamond 1983) suggests that visual feedback during development must be systematically correlated with self-initiated eye movements to play its role. In the absence of eye movements, visual feedback related to head, body, or limb movements is not appropriate for supporting formation of a representation of visual space. Other experiments by the same group, dealing with identifying the nature of the eye movement signals involved in visuomotor development, are reported in another section (see Chapter 4).

(b) Prism adaptation experiments

The above experiments pointed to the significance, for acquiring visuo-motor co-ordination, of the hierarchical status of the body segments with which action-related visual feedback is obtained. Further arguments supporting this point are provided by prism adaptation experiments in human adults.

The experimental design for studying prism adaptation follows a well-established procedure, typically involving three stages. The first stage is the pre-test. Pointing errors with one hand are measured during movements executed in the absence of visual feedback from the moving limb ('no visual feedback' situation). The second stage corresponds to exposure to the visuomotor conflict. The subject actively moves his hand (the exposed hand) under visual control, while wearing laterally displacing prisms. Finally, the third stage is the post-test. Pointing errors made by the exposed hand are measured in the 'no visual feedback' situation. The difference in pointing errors between the pre-test and the post-test (the after-effect) gives a measure of the degree of adaptation (see Held and Freedman 1963; Held 1965).

A clear example of hierarchical organization of visuomotor co-ordination is given by a series of prism experiments made by Prablanc and his colleagues. These experiments were aimed at manipulating the degree of generalization of adaptation from one exposed limb or segment to other, non-exposed, segments. Prablanc *et al.* (1975a) first replicated the classical finding that brief exposure (a few minutes) of one hand to the prism deviation (with head fixed) leads to an after-effect which remains localized to the exposed hand, and does not transfer to the other, unexposed, hand (Fig. 2.5). Lack of transfer of adaptation in this condition was further verified in another experiment where the two hands were successively and separately exposed to opposite prism deviations: the right hand was exposed to a rightward prism deviation through a prism placed in front of the right eye; immediately afterwards, the left hand was exposed to a leftward prism deviation through the left eye. During the post-test, after-effects of opposite signs and of approximately equal amplitude were found for each hand (Fig. 2.6 and Table 2.1). If transfer of adaptation between the two arms had taken place, after-effects of opposite signs should normally have cancelled each other, which was not the case because significant after-effects were present with both hands (Prablanc *et al.* 1975b). In a third experiment, the arms were not exposed to the prism deviation. Instead, the subject, with one eye wearing the prism and the other one occluded, attempted to reach targets placed in front of them with a rod rigidly held in his mouth. The task therefore only involved moving the head and the

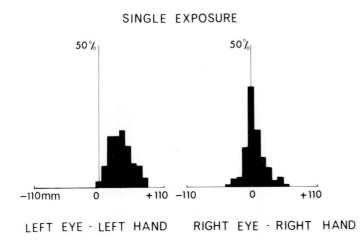

SINGLE EXPOSURE

LEFT EYE · LEFT HAND   RIGHT EYE · RIGHT HAND

Exposure		Open loop pointing	
LE	RE	LE–LH	RE–RH
◁	▬	+37 ***	+6

Fig. 2.5. Adaptive after-effect obtained by exposing one arm to a visuomotor conflict produced by a laterally displacing prism. A 17.5-dioptre leftward displacing prism was placed in front of the left eye in six subjects. The right eye was occluded. During the exposure phase (3 min) the subjects observed the movements of their left hand across their visual field. The right hand remained immobile and invisible. Following exposure, subjects pointed at visual targets seen monocularly with their left (exposed) hand and their right (unexposed) hand, in the 'no-visual feedback' condition.

The histogram on the left represents the distribution of pointing errors made with the exposed hand; zero corresponds to the location of the targets. As expected, pointings were deviated to the right, that is, in the direction opposite to the prism displacement. This after-effect did not transfer to the unexposed hand: as shown by the histogram on the right, pointing errors with the right hand were distributed around the location of the targets.

The table below the figure represents the exposure condition and the mean after-effects obtained with the exposed and the unexposed hand. xxx indicates an after-effect significantly different from zero at $P = 0.001$ (positive values: deviation of pointings to the right of targets; negative values, to the left. 110 mm represents the value of the optical displacement produced by a 17.5-dioptre prism). LE, RE, left, right eye; LH, RH, left, right hand. (From Prablanc *et al.* 1975a.)

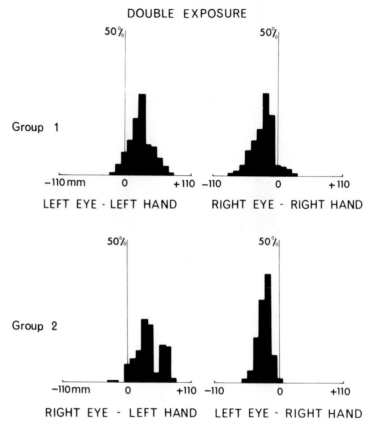

Fig. 2.6.   Adaptive after-effects obtained by exposing the two arms to visuomotor conflicts produced by two prisms displacing the visual field in opposite directions.

Two groups of subjects were exposed successively to a leftward displacement of their left hand, viewed through their left eye, and to a rightward displacement of their right hand, viewed through their right eye. The two groups differed only in the way the after-effects were tested. The subjects of group 1 were tested with the same eye–hand combination as during exposure (as in Fig. 2.5). Note opposite after-effects on each hand, both in the expected direction (see Table 2.1). Subjects of group 2 were tested during pointing with one hand at targets seen by the opposite eye. This procedure did not affect neither the direction nor the size of the after-effects (see Table 2.1), hence indicating that the after-effect is related to the program of the movements and not to a change in visual space. (From Prablanc *et al*. 1975a.)

trunk. When adaptation was obtained, after-effects were tested for each arm separately. A large after-effect was found on both sides (Table 2.2). This result (Prablanc, Tzavaras, and Jeannerod, unpublished) therefore demonstrates a proximo-distal polarization of the prism adaptation effect. Although adaptation was not generalized when exposure was limited to one hand, generalization did occur when exposure involved more proximal

Table 2.1

Group I	Subjects	Left eye–left arm	Right eye–right arm
	1	+25.9	−16.9
	2	+35.7	−34.7
	3	+51.9	−34.3
	4	+8.7	−22.4
	5	+28.5	−19.3
	6	+12.3	−18.79
Total mean		Mean 1 = + 27.2	Mean 2 = − 24.4
Variance		495	583
d.f.		286	286
t		10.36	8.58
P		<0.0001	<0.0001
Group II	**Subjects**	**Right eye–left arm**	**Left eye–right arm**
	7	+31.6	−32.9
	8	+56.9	−20.6
	9	+34.8	−20.6
	10	+19.8	−35.7
	11	+61.6	−24.0
	12	+13.4	−29.3
Total mean		Mean 3 = + 36.4	Mean 4 = − 27.2
Variance		538	294
d.f.		286	286
t		13.3	13.47
P		<0.0001	<0.0001

+, indicates a shift to the right.
−, indicates a shift to the left.
(From Prablanc *et al.* 1975b).

segments. This confirmed that visuomotor adaptation can be generalized from proximal to distal segments, but not in the reverse direction. A similar explanation holds for other results showing that if adaptation is obtained by effecting arm movements with a proximal joint (e.g. the shoulder), its effects tend to generalize to the more distal segments of the limb, although the reverse is not true: adaptation obtained by moving the wrist does not transfer to shoulder movements (Putterman *et al.* 1969; Hay and Brouchon 1972).

2.2. The co-ordination of proximal and distal components in prehension movements

Co-ordination of the different musculoskeletal segments that are involved in hand reaching and grasping is another important aspect in the organiz-

Table 2.2. Mean values of after-effects following attempts to reach targets with a rod held in the mouth, during exposure to a rightward displacing prism worn in front of the right eye (left eye was occluded). Following this exposure, the prism was removed and pointing errors were measured for both hands in the 'No-visual feedback' condition. An after-effect to the left was observed for both hands. Values of the after-effects (in mm) are different from zero at $P = 0.001$ (Student t.test).

RE–LH	RE–RH
-52^{xxx}	-48^{xxx}

ation of goal-directed movements. The different segments that compose the arm simultaneously subserve different parts of the same action. This aspect of arm movements is not apparent in pointing, where the whole hand is merely transported towards the target by rotation of only the proximal joints. By contrast, in more complex actions like prehension, which involve reaching to an object and grasping it, the hand assumes movements and postures that are apparently independent of those assumed by the more proximal segment of the limb.

Formation of the finger grip during the action of grasping a visible object involves two main functional requirements, the fulfilment of which will determine the quality of the grasp. First, the grip must be adapted to the size, shape, and use of the object to be grasped. Second, the relative timing of the finger movements must be co-ordinated with that of the other component of prehension by which the hand is transported at the spatial location of the object. In particular, closure of the finger grip must occur in tight synchrony with approach of the target object with the fingertips. Both early and late closure will result in an inaccurate grasp, with the consequence of bumping and eventually breaking fragile objects. Thus, a study of grip formation during prehension has to take into account, not only the motor pattern in itself as it is obtained by discrete finger posturing, but also its change over time until prehension is actually achieved.

Simple observation of prehension movements shows that finger posturing anticipates the real grasp and occurs during transportation of the hand. This hand 'shaping' (of which grip formation is the most representative aspect) is therefore related to purely visuomotor mechanisms, that is, mechanisms that are independent from manipulation itself. Manipulation, which occurs after the object has been grasped, relates to co-ordination between finger movements and tactile and kinesthetic inputs: this aspect will not be considered in the present context. Visually guided prehension movements have received relatively little attention, partly because the large number of degrees of freedom involved in such movements makes their experimental approach difficult. Until recently only global descriptions have been given, based on presence or absence of grip formation, or on time taken to perform a grasping task.

Duality of visuomotor mechanisms in prehension reflects in part the organization of sensory systems. Although objects are perceived as phenomenal entities, sensory systems are known to detect features, not objects. Objects have to be split into elemental visual features, or properties, like size, shape or texture, each of which is assumed to activate a specific visual mechanism. Such 'intrinsic' properties are constituent of object identity. In addition, when object perception is considered in the behavioural context, another set of properties emerges. Objects have a specific orientation, distance from the body, and location in the frontal plane—these are their 'extrinsic' properties.

Both 'intrinsic' and 'extrinsic' properties of objects are essential attributes for governing actions directed toward them, especially if one assumes that the different properties of an object are matched by specific mechanisms that generate motor commands appropriate for each property (see Paillard and Beaubaton 1974; Jeannerod 1981a, b; Jeannerod and Biguer 1982). These mechanisms can be conceived as specialized input–output structures, or visuomotor channels, which function simultaneously for extracting a limited number of parameters from the visual world and for producing the corresponding responses. The notion of parallel visuomotor channels governing the two components of prehension is in fact complementary to the other notions of a single act and a unified percept. One can further speculate that visuomotor channels only represent selective pathways for the input–output information flow related to each component; but that the movement as a whole is represented by a unique programme governing the integrated aspect of the action, or in other words the co-ordination of the components. Accordingly, the musculo-skeletal segments related to the act of prehension, in addition to their differential involvement in independent channels, could also be constrained as a motor ensemble (according to the concept of co-ordinative structure, see Chapter 1). The co-ordinative structure related to prehension would be governed by a specific set of rules, hierarchically higher than those of the channels, and co-ordinating the activity of the channels in the time domain. Interactions between the two components of prehension have been conceptualized by Arbib (1981) in a model which stresses both separate activation of each component by specific visual pathways, and co-ordinated output (Fig. 2.7). The notion of uniqueness of the program is thus in theory not incompatible with that of parallel visuomotor channels. Some of the implications and recent developments of this theory have been reviewed by Marshall (1984).

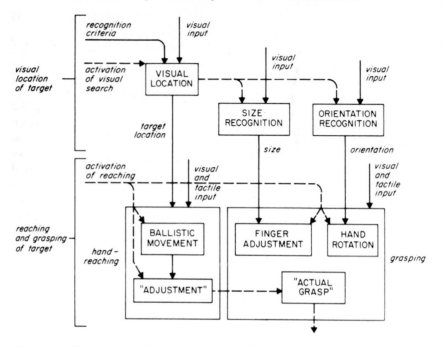

Fig. 2.7.   The Arbib model for accounting for independence of segmental components in reaching. Note parallel visual processors dealing with the various aspects of the object to be grasped and manipulated (spatial location, size, orientation). These processors are connected with controllers for the corresponding segmental movements (ballistic movement, finger adjustment, hand rotation). (From Arbib 1981, with permission.)

2.2.1. A study of prehension in normal adult subjects

(a) Methods

This study was based on the study of films of prehension movements directed at three-dimensional graspable objects. The degree of visual feedback from the moving hand during prehension was controlled by way of an apparatus consisting of a box resting on a table, divided horizontally into two equal compartments by a semi-reflecting mirror (Fig. 2.8, insert). Subjects were seated in front of the box with their forehead resting on the front panel. They looked through a window within the upper compartment and placed the hand under study into the lower compartment. Target objects were placed in the lower compartment along the subjects' sagittal axis. These were small solid objects, such as a sphere, a cube, or a vertical rod. Distance from the body could be varied (e.g. 25, 32, or 40 cm). Two experimental situations were used. In the control situation ('visual feedback' condition), subjects could see the lower compartment through the mirror

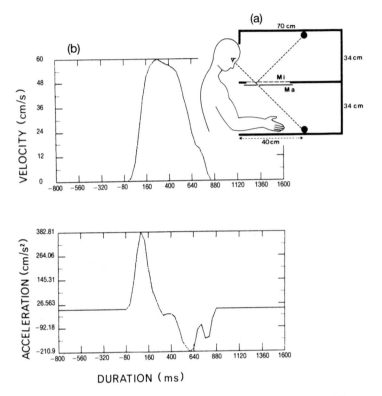

Fig. 2.8. (a) Apparatus for measuring prehension movements. Mi: two-way mirror. Ma: mask. Condition represented here is 'no visual feedback' condition.

(b) Velocity and acceleration profiles of arm during a single prehension movement reconstructed from films at 50 frames s^{-1}. Target placed at 32 cm from body. 'No visual feedback' condition. Total movement duration: 800 ms. Time to velocity peak: 280 ms. Onset of reacceleration: 600 ms. Curves have been smoothed by using a least-square polynomial approximation. Frequency cut-off: 5 Hz. (From Jeannerod 1984, 1986a.)

and therefore, they could see both the target object and their moving hand. In the other situation ('no visual feedback' condition), vision of the hand was prevented by inserting a mask below the mirror, so that the lower compartment was no longer visible. In the latter situation, target objects had to be displayed from the top of the upper compartment (Fig. 2.8, insert). Since the mirror was placed half-way between the target display and the table, subjects saw in the mirror a virtual image of the object, projected at table level. Another object identical to that seen in the mirror was placed directly on the table in exact coincidence with the virtual image. Thus, subjects reached for the virtual object below the mirror without seeing their hand, and met the second, real, object at the expected location.

During the experiment, subjects had to place the hand under study on a starting block near the body axis, with the forearm in the prone position and the fingers semi-flexed. They were required to perform rapid and accurate movements, to grasp the target-object as precisely as possible, and to carry it near the starting block. No formal time constraint was given. At the beginning of each trial a new object was displayed, while the subjects kept their eyes closed. At an acoustic signal they opened their eyes, and had to wait 2–5 s until a small light was turned on in front of them before performing the reaching movement.

The radial aspect of the subject's hand was filmed with a cine-camera running at 50 frames/s. Data were processed by projecting frame by frame the image of the movement on a screen with a one to one magnification. Duration of the movement was measured as the number of frames between the first detectable arm displacement and contact with the target object. Position of anatomical details on the wrist was plotted over successive frames. Distance between successive positions gave a measure of the instantaneous tangential velocity of the arm trajectory. From the same frames, the relative positions of the tip of the index finger and the tip of the thumb were also plotted. This allowed measurement of the size of the finger grip and its change over time. Due to the resting posture of the hand and the shape of the objects, no rotation of the wrist occurred during the movement.

The following description is drawn from a study of prehension in seven young adults (Jeannerod 1984). In this and the following sections, two components will be described for prehension movements. The trajectory of the arm between the starting position and the object will be referred to as the 'transportation component'. Formation of the grip by combined movements of the thumb and the index finger occurring during the arm movement will be referred to as the 'manipulation component'.

(b) Transportation component

This component was found to be little affected by whether visual feedback from the moving limb was present or not. In the 'no visual feedback' condition, movement duration was found to vary across subjects (average values between 674 ms and 1013 ms for a target located 40 cm away from the body), although it remained relatively constant for each given subject (coefficients of variation were within 10 per cent in most subjects). The general pattern of the transportation component was that of a reverted U-shaped trajectory. The hand was first raised from the support and then lowered down to the object. Its tangential velocity profile involved a fast rise of velocity up to a peak, which was reached at an average value of 308 ms from movement onset. In the three subjects where this was tested, the amplitude of the peak velocity was found to increase linearly with target distance, although movement duration remained relatively invariant.

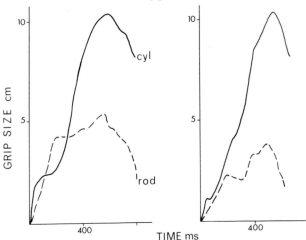

Fig. 2.9. Change in grip size (cm) over time (ms) in two different subjects reaching either for a small object (a 2-mm dia., 10-cm long rod) or a large object (a 55-mm dia., 10-cm long cylinder). Dashed lines correspond to movements aiming at the rod, solid lines at the cylinder. Note different grip sizes, but similar curve shapes. (From Jeannerod 1981b.)

Deceleration of the arm trajectory was marked by a discontinuity where tangential velocity tended to become constant or even to increase again before the movement was stopped (Fig. 2.8 upper diagram). Although the peak on the acceleration graph corresponding to this discontinuity (Fig. 2.8, lower diagram) rarely reached positive values of acceleration, it was nevertheless likely to represent re-acceleration of the arm partly damped by inertia of the limb. The time of occurrence of the discontinuity (measured at the lowest point on the acceleration graph) consistently corresponded to 70–80 per cent of the total movement duration.

(c) Manipulation component

Grip formation took place during transportation of the hand at the object location and was clearly anticipatory with respect to the grasp. In the experimental situation described here, the resting position imposed on the subject's hand before performing the movements implied semi-flexion of the fingers. As the arm was displaced, the fingers began to stretch and the grip size increased rapidly up to a maximum. After this time, the fingers were flexed again and the grip size was reduced to match the size of the object. The size of the maximum grip aperture was proportional to the anticipated size of the object. An example of this behaviour is given in Fig. 2.9, where grip size is compared in two subjects (Subjects 1 and 2 of Jeannerod 1984) during movements aiming at objects of different sizes (see also Table 2.3).

Table 2.3. Mean value (in mm) of maximum and final grip size (before contact) as a function of object size and visual feedback condition, in two subjects.

	Rod		Cylinder	
	Visual feedback	No visual feedback	Visual feedback	No visual feedback
Maximum grip size	31.9(4.7)	34.7(7.7)	74.6(3.0)	80.2(4.3)
Final grip size	13.3(1.9)	12.6(2.9)	58.1(2.4)	57.7(6.3)

Numbers in parenthesis are SDs.
(From Jeannerod 1986)

The biphasic pattern of grip formation (extension followed by flexion of the fingers) was confirmed by Wing and Fraser (1983) and Manchester (1985). These authors observed in addition that changes in grip size during prehension were effected almost exclusively by finger excursion while the thumb tended to keep an invariant position throughout the movement.

Finally, a similar sequence of anticipatory finger movements has been shown to occur during the action of catching moving objects (e.g. a ball). According to Alderson *et al.* (1974), closure of the fingers begins prior to contact of the ball with the palm of the hand.

(d) The pattern of co-ordination between the two components

Although transportation and manipulation appear to be organized independently from each other, they share a common time course. This fact, which is essential for understanding the co-ordination of the two components, is reflected by synchronous changes of the kinematics of the two trajectories. In the present experiment, it was found that the time occurrences of the re-acceleration of the arm and of the maximum grip size were strongly correlated to each other. Figure 2.10 demonstrates this relationship in one subject. By superimposing the averaged profiles of arm acceleration and of grip size for the same prehension movements as a function of time, it becomes clear that the maximal grip size closely corresponded to the beginning of the arm re-acceleration (Fig. 2.10, left). This point was confirmed by a regression analysis of the time occurrences of arm re-accelerations with respect to those of maximum grip sizes (Fig. 2.10, right).

Concerning this point, Fraser and Wing (1981) reported an interesting anecdote in recording grasping movements in one subject equipped with an artificial arm. They observed that the time relation between the arm trajectory and grip formation was preserved with the artificial hand to about the same extent as with the other, normal, hand. This co-ordination, however, was effected in a strikingly different way for each hand. The subject oper-

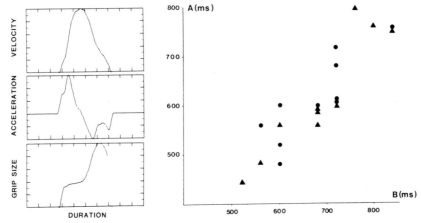

Fig. 2.10. Time relation between the two components of prehension. Left: averaged velocity and acceleration profiles of 20 transportation components in one subject, and of the grip size of the 20 corresponding manipulation components. No visual feedback condition. Right: plot of the time occurrence of maximum grip size (B, ms) versus onset of re-acceleration of transportation component (A, ms) for the same movements. Correlation between A and B was $r = 0.81$, $P < 0.001$. Data from subject 7 of Jeannerod (1984).

ated his artificial hand by a shoulder harness and hand opening was obtained by shoulder protraction. This observation suggests that the co-ordination between the two components of prehension is not the effect of biomechanical coupling between arm and finger movements, but rather pertains to a centrally generated temporal structure, which can be generalized to other, equivalent, object-oriented plurisegmental movements.

Indeed, the same type of co-ordination of two components has been demonstrated in an animal analogue, the action of pecking in birds. Grasping seeds with the beak involves co-ordination between the head movement (pecking) and changes in the gape size. The systematic film study of Klein *et al.* (1985) in the pigeon revealed striking similarities to human prehension. In both avian pecking and human reaching the velocity profiles of the transportation component have similar shapes and involve a similar relationship of velocity to distance from the target. In both cases also the apertures of the prehensile organ (jaw or hand) are sized to the target, and their kinematics are related to those of the transportation component. Finally, as in humans, vision during the movement does not seem to be required for grasping in birds: according to Klein *et al.* (1985), pigeons begin to close their eyes when they start their pecking movement.

Neural mechanisms for head–neck and for jaw movements in birds are segregated in the same way as are mechanisms for axial–proximal and distal movements in primates (see below). In the case of birds the motor neurons controlling axial and distal components of the musculature are anatomically separated. Spinal neck motor neurons are located medially (Eden and Correia 1982), although brainstem jaw motor neurons are located laterally (Wild and Zeigler 1980). This anatomical disposition might correspond to the involvement of different descending motor pathways for the control of the different movement subtypes.

(e) Degree of independence of the two components

Assuming that the two components of prehension are generated independently with respect to each other implies that the motor program of a given component can be modified without affecting the motor program of the other.

In an attempt to demonstrate such a dissociation, Jeannerod (1981b) recorded prehension movements during which the size and shape of the object to be grasped was unexpectedly changed. This effect was produced in using the apparatus described in Fig. 2.8. An ellipsoid object (7×4 cm) was placed above the mirror. This object was mounted on a support in such a way that it could rotate 90° in a plane perpendicular to the mirror plane. According to the orientation of the object, the subject saw in the mirror either a 7-cm long ellipsoid object or a 4-cm diameter spherical object. Instantaneous rotation was produced by a small motor triggered by raising the hand from the starting position. Each trial was started with the presentation of a sphere, and transformation into an elongated shape occurred randomly across trials. The subjective appearance was that of a sudden expansion of the object shape.

The transportation component was not affected by the changing shape of the target object at the onset of the movement. Duration, maximum velocity, and location in time of the re-acceleration remained within the same range in both the normal trials and those involving a change in object shape ('perturbed' trials). The pattern of the grip, however, differed according to whether the object shape remained constant or was unexpectedly altered. This difference was particularly consistent in the subject illustrated in Fig. 2.11. The time at which a difference in hand shaping between the two conditions became visible can be grossly evaluated by comparing the finger postures during one normal (Fig. 2.11(a)) and one perturbed trial (Fig. 2.11(b)). Although 420 ms after the perturbation had occurred, the finger postures were still similar in the two movements, a clear difference became visible at a later stage. This result suggests that at least 500 ms are needed for processing the visual information related to the change in shape and for elaborating the new motor command. If one assumes that the finger posture must be altered before the fingers begin to close (i.e., some 200 ms before the end of the movement), it appears that the possibility of such an alteration occurring depends critically on the movement duration. Reaching movements shorter than about 700 ms would thus not allow enough time for finger posture to adapt to such a change.

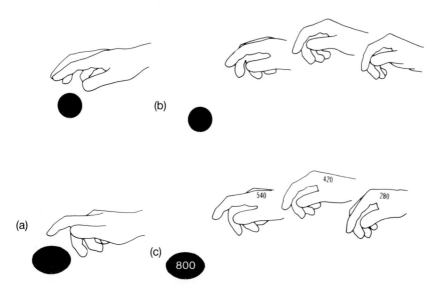

Fig. 2.11. Differential hand shaping in visually perturbed and non-perturbed trials, in one subject. (a) Pattern of anticipatory hand shaping in movements directed at a sphere (non-perturbed trials) or at an ellipsoid shape (perturbed trials). (b) Non-perturbed trials. (c) Perturbed trials. Hand posture has been drawn at corresponding times of the movements (i.e. 280, 420, and 540 ms after the occurrence of the perturbation). Both movements have the same total duration (800 ms). Notice the beginning of a detectable differential shaping on the frame at 540 ms. (From Jeannerod 1981b).

Another attempt at dissociating the two components of prehension was made by Wing *et al.* (1986). Their subjects were requested to perform reaching movements at different speeds. Duration of movements performed at a 'normal' speed averaged 735 ms, that of movements performed at 'fast' speed, 376 ms. Measurement of finger grip size as a function of speed of the movement revealed that maximum grip size associated with fast movements was on average 13 mm larger than with normal movements. In addition, in movements performed at normal speed, but without vision of the hand, maximum grip size was on average 35 mm larger than in the same movements performed under visual control. Wing *et al.*'s interpretation of these results was that exaggerated grip aperture occurred in conditions where accuracy of the transportation component was degraded (e.g. fast speed, lack of visual control), the function of the enlarged grip being to increase tolerance for terminal errors of the hand. This interpretation (based on Fitts' law; see Chapter 3) has important implications, in suggesting that the manipulation component is determined, not only by the physical dimensions of the object, but also by the

dynamic aspects and the accuracy constraints of the whole action of prehension.

(f)   The role of visual feedback in prehension movements

The main role of visual feedback in prehension seems to be to achieve terminal accuracy. Wing and Fraser (1983) suggested that the relative stability
of the thumb position with respect to the wrist during grip formation might
provide a landmark of the end-point of the limb. The regulatory action of
visual feedback on movement dynamics would be exerted through this
landmark, in order to adjust the amplitude of the transportation component to the precise location of the target object. This mechanism would be
consistent with what is known about the role of visual feedback in terminal
guidance of the movement (see Chapter 3).

The action of visual feedback on prehension can be determined by comparing movements executed with or without vision. Jeannerod (1984)
found that prehension movements executed in the 'visual feedback' condition tended to have an overall longer duration than those executed in the
absence of visual feedback. This difference in duration was not evenly distributed throughout the trajectory of the movements. The duration of the
first submovement (i.e. up to the re-acceleration) was shorter in the 'visual
feedback' condition. As a consequence, the duration of the second submovement (after the re-acceleration) was significantly longer when visual
feedback was present. This result indicates that visual feedback was incor-
porated into the pre-existing structure of the movements, the additional
time needed to process visual feedback thus explaining the overall increase
in movement duration and the increase in duration of the second submovement.

Apart from this role in achieving terminal accuracy, visual feedback
seemed to have a limited role in maintaining the temporo-spatial pattern of
prehension. Specifically, re-accelerations observed during the deceleration
phases were still present in the 'no visual feedback' situation, which
excludes a purely visual corrective mechanism for these re-accelerations.
They should rather be looked at as an effect of central patterning of motor
output. For instance, they might represent 'positioning' movements
(Annett *et al.* 1958) or target 'acquisition' phases (Welford 1968) related to
constraints appearing in movements involving several segments and requiring a high degree of precision.

Similarly, presence or absence of visual feedback did not affect the manipulation components. Complete and accurate grip formation was observed
in the absence of visual feedback from the hand. Jeannerod (1981b, 1984)
found no significant difference in the values of maximum grip size and final
grip size before contact with the object between the 'visual feedback' and
the 'no visual feedback' conditions (see Table 2.3). This finding is in apparent contradiction to those of other authors who found that the amplitude of

finger excursion and the value of maximum grip size were reduced (Manchester 1985) or increased (Wing *et al*. 1986) in the absence of visual feedback from the moving limb. Experimental conditions might account for these differences.

Finally, the relative synchrony of arm re-acceleration with finger closure, as observed in the above results, is also evocative of a central co-ordination. Control by sensory feedback may co-exist with the central mechanisms. However, co-ordination and corrections based on sensory feedback (visual or proprioceptive) should require positional signals from limb position with respect to the target, limb position with respect to the body and respective segmental positions. These signals were not available from the visual modality in our experiment with the 'no visual feedback' condition and therefore, it is doubtful whether vision would be an appropriate controller for segmental positions. It remains to be demonstrated whether proprioception would effectively represent at the central level a source of positional signals for intersegmental co-ordination. This point will be discussed in Chapter 5.

2.2.2. Development of prehension movements

Another justification for considering the proximal and distal components of prehension as separate (although co-ordinated) entities can be drawn from behavioural studies of the development of prehension in human infants and in young animals.

In babies, these studies have revealed a synchronous development of the two components. Finger posturing is lacking during reaching for visual objects until the age of approximately 20 weeks. According to Halverson (1931), inaccurate posturing is then observed, but it is not until the age of 36–52 weeks that a precision grip can be formed (von Hofsten and Fazel-Zandy 1984; Lockman *et al*. 1985). Other studies, however, have shown that infants are able to make use of visual information for crude finger posturing earlier than previously suspected. According to Bruner and Koslowski (1972), infants 10–22 weeks of age may show coarsely-adapted hand movements when they are presented with small graspable visual objects within reach. This is not the case for objects of a larger size, exceeding the grasping capability of the hand. Von Hofsten (1982) has shown that even younger infants (1 week) may intercept the trajectory of moving objects and come into contact with them. These movements are at first jerky and oscillatory, but they improve rapidly up to a point where two phases can be discerned, namely, an initial rapid arm extension lasting around 500 ms and then a series of stepwise smaller movements (von Hofsten 1979). These reaching movements, however, are effected with a widely open hand without evidence of grip formation. It is only when the object has been touched that crude prehensile movements of the fingers can be observed.

The late occurrence of hand shaping is controversial; some babies are reputed to show it within the first weeks of life (Bower *et al.* 1970). Their movements, however, are of an immature style. They do not seem to be related to the presence of a visual object (Bower *et al.* 1970; Trevarthen 1982). In addition, although maximum finger opening seems to be related to maximum extension of the wrist (hence suggesting a co-ordinated reach and grasp pattern), they are much slower than adult movements. Trevarthen has suggested that finger movements observed in very young infants might represent the expression of a pre-formed proximo-distal motor pattern (pre-reaching), which later will evolve into mature prehension (see Trevarthen 1984). Observations by T. Humphrey (1969) suggest that crude reach and grasp movements might even be present in human foetuses at an age when it may be assumed that only spinal motor centres are functional.

One can hardly determine whether poor ability to adapt finger posture to the shape of a visual object in human infants is due to incomplete maturation at the input or the output levels of the visuomotor system that controls the movement. In favour of the input level, DiFranco *et al.* (1978) observed that an infant at the pre-reaching stage will reach as actively for a two-dimensional picture of an object as for the real object. This observation, however, seems somewhat contradictory to that of Bruner and Koslowski (1972, see above) showing that babies will attempt to reach only for objects of a graspable size. As for the output level, it is worth stressing the deleterious effects on hand shaping of early reflex reactions and synergies. In Twitchell's (1970) terms, the 'initial prehension appears more accidental than intentional' and 'the emergence of the instinctive avoiding response at the time of these early attempts at voluntary prehension causes ataxia of reach and overpronation of the hand . . . '. Twitchell (1970) recognized that dexterous prehension appears late and that small objects remain difficult for the infant to handle; 'the required thumb–fingers apposition does not appear until the grasp reflex can be fractionated during the second half-year of life'. Later, 'the instinctive avoiding response can contaminate activity and the fingers may abduct or dorsiflex too much as the hand is extended toward the object'.

In order to discuss other possible justifications for the concept of a duality of motor mechanisms in prehension, it is tempting to relate the developmental time course of finger movements in children to the maturation of motor pathways. Accordingly, the late development of finger movements and the persistence of instinctive reactions might reflect immaturity of cortico-motoneuronal synapses controlling independent hand and finger movements. It is known that pyramidal fibres continue to increase in diameter up to somatic maturity. In addition, it has been shown that the pyramidal tract myelinates relatively late in humans: myelination increases up to the eighth post-natal month, and seems to be complete around the age of one year (Yakovlev and Lecours 1967) or even two years (Lang-worthy 1933). These results indicate that consistent formation of a precision grip during prehension of visual objects would be contemporary with the existence of a functional corticospinal tract. In humans, however, more data are still needed for establishing a precise anatomic–functional correlation.

As in children, very young monkeys develop goal-directed arm extension earlier than manipulative movements. Recent observations by J. Vauclair (personal communication, 1986) in the baboon showed that, during the first three post-natal weeks, the animal makes tentative unimanual reaches directed at visual objects (e.g. pieces of food). In these reaches, either hand is thrown in the direction of the food target with the fingers fully extended (unspecified reaches, 75 per cent of cases) and eventually achieves a palmar grasp (25 per cent of cases). Palmar grasp develops at the expense of unspecified reaches, but precision grip does not appear until the seventh week after birth. Interestingly, it is only at the time when precision grip appears, that hand preference becomes detectable and that unimanual reaches become clearly lateralized.

In the monkey, dissociation between development of proximal and distal components of prehension does reflect immaturity of cortico-motoneuro-nal synapses that control independent hand and finger movements. According to Kuypers (1962), this pathway does not fully develop until the eighth month in macaques, and it is not until then that these animals can make a precision grip.

2.2.3. Effects of brain lesions on prehension movements

Mechanisms responsible for grip formation and those that account for carrying the hand to the object location can also be dissociated by localized brain lesions occurring in pathological conditions or created experimentally.

(a) Lesion of motor corex and pyramidal tract in the monkey

One of the most generally accepted facts from experimental studies of pyramidal function in monkeys is the critical importance of the corticospinal tract in the control of discrete finger movements. This notion was first introduced by Tower (1940) and later confirmed by Lawrence and Kuypers (1968) and Woolsey *et al.* (1972), who all showed that section of the pyramidal tract produces a long-lasting, if not permanent, inability to perform independent finger movements, as in grasping small pieces of food. The classical observations made by Tower (1940) following unilateral pyramidotomy illustrate this point. Tower had noted that 'all discrete usage of the digits is utterly and permanently abolished', especially in those actions where 'the digits are used separately as in the opposition of thumb and index to pick up small objects, or in fine manipulation of objects, or in separation of one digit from the others . . . ' (p. 55). This specific deficit was well illustrated by inability of the animals to execute grooming or by their awkward attempts to grasp food.

Instead of the normal movement which culminates in the opposition of thumb and index to pick up small objects, . . . the residual movement of the paretic arm is a highly stereotyped reaching–grasping act involving the entire body half, similar to the reaching–grasping act of the newborn monkey. The hand is brought down on the object in half pronation and scoops it into the ulnar side of the hand, unless, as frequently happens, the fist closes before reaching the object. (Tower 1940, p. 56)

The degree of recovery from this deficit has been quantified by Chapman and Wiesendanger (1982) in monkeys, following unilateral section of the pyramidal tract at the pontine level. The animals' performance was scored in a Kluver board task, i.e. the number of pellets of food extracted from holes of different sizes and the time needed for clearing the board, were measured for each hand. The strategy utilized by the animals for grasping the pellets was found to be changed after pyramidotomy when the hand contralateral to the lesion was tested. Animals first used all four fingers for extracting the food, a strategy that was efficient only for the larger holes. At a later stage (30–40 days after surgery), monkeys regained the ability to pry out the pellets from the smaller holes by using either the thumb or the index finger. Once extracted, the pellets could be grasped between the opposed thumb and index finger, but this precision grip remained weak and clumsy and the food was eventually dropped before reaching the mouth. Partial recovery of grip formation in the Chapman and Wiesendanger experiment might be explained by incompleteness of the pyramidal tract lesions.

Neonatal lesion of the corticospinal system corresponding to the hand area results in lack of development of the distal component of prehension. Lawrence and Hopkins (1972) and Passingham *et al.* (1978) have shown that monkeys on which a complete unilateral ablation of area 4 was performed during infancy never acquire a precision grip when tested later in adulthood. As previously mentioned, precision grip in monkeys normally develops around the eighth post-natal month, i.e. by the time when maturation of the corticospinal tract is completed, as judged from the formation of cortico-motoneuronal synapses (Kuypers 1962) and the level of excitability of motor neurones by motor cortex stimulation (Felix and Wiesendanger 1971). Effects of cortical or corticospinal lesions on finger movements in monkeys thus confirm the evidence accumulated in anatomical and physiological experiments for the predominant cortical involvement in the control of finger muscles [e.g. Phillips and Porter (1977). For a recent confirmation of these findings, see Muir and Lemon (1983), Muir (1985)]. A similar cortical dependence of distal movements was recently demonstrated in the cat. Burgess and Villablanca (1986) showed that neo-natal hemispherectomy impaired the maturation of the normal food pre-hension response. Animals tended to throw their affected limb in the direction of the food target, with a stiff and clubfooted jab, instead of the typical prehensile-like posture (see also Armand *et al.* 1986). This deficit, however, was less severe in neonatally lesioned animals than in adult-lesioned animals.

(b) Pyramidal lesion in humans

Lesions involving motor cortical areas or their efferent pathways (as in hemiplegia following stroke, for instance) are of relatively frequent occur-rence. In studying recovery from hemiplegia in patients with surgical resec-tion of area 4, Hécaen and Ajuriaguerra (1948) noticed that movements of the proximal joints recovered first, with a return to normal muscular force within 4–6 weeks. By contrast, fine and isolated finger movements appeared to be permanently lost in these patients. Similarly, Lough *et al.* (1984) confirmed that patients with hemiplegia usually recover the shoul-der–elbow synergy for transporting the hand near an object, provided the shoulder is passively supported against gravity, but that finger movements seem to remain indefinitely clumsy: during prehension, they do not shape in anticipation to the grasp, which is achieved with the palmar surface of the whole hand instead of the fingertips. Besides adult hemiplegia, there are other pathological conditions where a specific alteration of finger movements can be studied more easily. One of these conditions is infant hemiplegia.

Infant hemiplegia is a disease of unknown origin associated with malfor-mation and/or lack of maturation at the cortical level within one hemi-sphere. It consists in spastic palsy of the limbs on one side, sometimes accompanied by mild mental retardation. Interestingly, the hemiplegia is usually not noticed until the age of about 40 weeks, i.e., around the time where the hand becomes normally engaged in prehensile activities. At this early stage, the only noticeable deficit is non-use of the affected hand in manipulation normally requiring both hands. Spasticity may appear at a later stage (see Goutières *et al.* 1972 for review).

Finger movements during prehension have been examined by Jeannerod (1986) in two such patients aged 23 months and 5 years, respectively. In the younger patient (patient Mag), the affected (right) hand remained sponta-neously unused. It was only when the normal hand was immobilized that the right hand could be teased, though with difficulty, to grasp objects. Figure 2.12, redrawn from film records shows comparative records of grasping of a prong from a pegboard with either hand, in the 'visual feed-back' condition. The normal hand (Fig. 2.12(a)) appeared to shape incom-pletely with respect to the object, though the finger extension–flexion pattern was none the less clearly present. In addition, contact of the hand

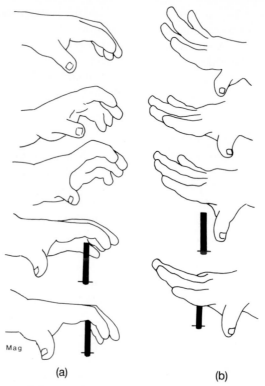

Fig. 2.12.   Pattern of finger grip in patient Mag, during reaching in the 'visual feedback' condition. (a) normal hand. (b) affected hand. Redrawn from film. (From Jeannerod 1986).

with the object triggered an immediate posturing of the fingers which ensured accurate grasping. By contrast, the affected hand remained exaggerately stretched throughout the duration of the movement, without any evidence of grip formation (Fig. 2.12(b)). Some posturing of the fingers occurred after contact with the object, resulting in a very incorrect and clumsy grasp.

In the other child (patient Gis), the affected hand was used spontaneously, partly as a result of training and rehabilitation procedures continued for several years. Better co-operation of this patient allowed more complete analysis of her prehension movements. With her normal hand, she performed correct and accurate grips with a fully developed, adult-like pattern. This point is illustrated by the example given in Fig. 2.13(a), and by the kinematic analysis of another movement represented in Fig. 2.14(a). Prehension with the affected hand differed from that of the normal hand only for what concerned the pattern of grip formation. In the three examples shown in Fig. 13 (b, c, and d), finger posturing appears

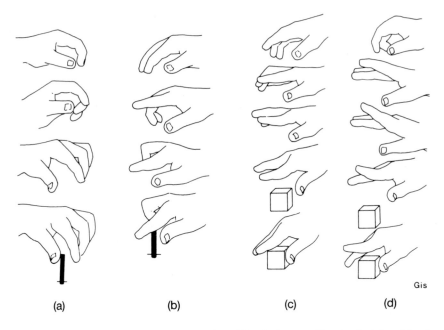

Fig. 2.13. Pattern of finger grip in patient Gis, during reaching in the 'visual feed-back' condition (a) normal hand. (b, c, d) affected hand. Redrawn from film. (From Jeannerod 1986.)

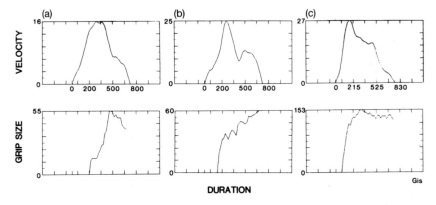

Fig. 2.14. Change in velocity of the transportation component and grip size of the manipulation component in individual prehension movements as a function of movement duration in patient Gis. (a) normal hand. (b, c) affected hand. 'Visual feedback' condition. Maximum grip size: (A), 55 mm; (B), 60 mm; (C) 153 mm. (From Jeannerod 1986).

abnormal. The index finger is exaggeratedly extended and flexes incompletely, if at all, before contact with the object. In the two other examples analysed graphically in Fig. 14 (b and c), no finger grip formation can be detected, although the velocity profiles corresponding to the transportation components appear relatively similar to those of the normal hand (Fig. 14(a)). The lack, or the abnormal character of grip formation with the affected hand resulted in awkward and clumsy grasps of the objects which were occasionally dropped from the hand.

(c) Lesion of the non-pre-central cortical areas

In addition to the pre-central cortex, higher order 'motor' cortical areas, such as pre-motor areas in the frontal lobe or posterior parietal areas, are also involved in the control of visually guided finger movements. Effects of lesions of the parietal cortex, which are entirely relevant to the present study, will be reported extensively in a special section (see Chapter 6). A survey of the prefrontal mechanisms for the movement control can be found in Humphrey (1979).

2.3. Possible dissociation of proximal and distal motor mechanisms in actions involving intra-hemispheric and inter-hemispheric processing

Visuomotor mechanisms in reaching involve collaborative action of the two hemispheres, especially in situations where one hand is used to respond to stimuli appearing within the contralateral visual field. In such situations it is necessary, for motor cortical regions in the hemisphere controlling the hand to be activated, that visual information can transfer from the other hemisphere. If this inter-hemispheric transfer were made impossible (by transsection of the corpus callosum, for example), the observed motor response to the contralateral visual stimulus would reflect only those aspects of visuomotor processing which can be effected within the hemisphere ipsilateral to the responding arm. Situations involving disconnection between the two hemispheres have been abundantly exploited for testing the respective roles of crossed and uncrossed motor pathways in controlling different types of movements. In addition, attempts have been made in normal subjects, by using the same rationale, to dissociate these two motor components.

2.3.1. Inter-hemispheric disconnection

Subjects with surgical section of the corpus callosum or lesions interrupting callosal transfer, show typical impairments in finger movements when they are properly examined. Gazzaniga *et al.* (1967) have examined finger

movements in split-brain patients. The patients were requested to reproduce hand and finger postures presented as outlines flashed on one side of
the visual field, so that only one hemisphere was stimulated at each presentation. They could easily reproduce finger postures with their hand contralateral to the stimulated hemisphere, although their ipsilateral hand usually
failed, except for the simplest postures. This result was interpreted by Gazzaniga and his co-workers (1967) as a demonstration of the fact that ipsilateral motor control is at its worst in tasks in which the hemisphere is
required to direct individual movements of the fingers.

Similar results have been obtained from animals. Monkeys with complete split-brain (i.e. involving also a section of the optic chiasm) are able
to efficiently intercept moving objects with either arm when vision is
restricted to one eye (and therefore, to one hemisphere). By contrast, they
can orient and shape their fingers according to the object size only with the
hand on the side opposite to the stimulated eye (and hemisphere). This
result (Trevarthen 1965) indicates that the visuomotor apparatus for reaching remains undivided by the split, although the visuomotor apparatus controlling grasping governs each hand independently. One possible
interpretation for such a dissociation is that visuomotor control of fine
finger movements can be effected only via the motor cortex and the crossed
corticospinal pathway. By contrast, more proximal movements can be
effected not only by the crossed pathway, but also by a more diffuse cortico-subcortical uncrossed route. Trevarthen (1965; see also Trevarthen
and Sperry 1973) thus proposed the hypothesis of a cortical visuomotor
control of fingers opposed to a subcortical (brainstem) visuomotor control
of proximal segments.

The dichotomy between levels of visuomotor control for proximal and
distal components of prehension has been confirmed by Lund *et al.* (1970)
and by Brinkman and Kuypers (1973), also in experiments with complete
split-brain monkeys. In the Brinkman and Kuypers experiments, animals
were tested monocularly with an improved version of the Kluver board,
where information necessary for extraction of food pellets was restricted to
visual cues only (i.e. tactile cues were eliminated). Extraction of the
pellets, which required formation of a precision finger grip, was possible
only with the hand contralateral to the open eye. The other hand could
reach to the board, explore it tactually, but not shape according to the
visual aspect of the food target (Fig. 2.15). In another series of experiments
also with split-brain monkeys, Haaxma and Kuypers (1975) used a more
refined food target, such that extraction of the food pellets required not
only precision grip formation but also correct orientation of the thumb and
index fingers. They showed that proper control of the hand contralateral to
the open eye became inefficient if a deep parieto-occipital leucotomy was
made within the corresponding hemisphere. Haaxma and Kuypers interpreted their results as an effect of intra-hemispheric disconnection between

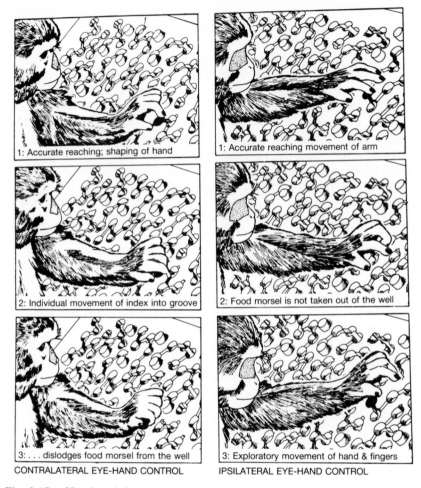

Fig. 2.15. Hand and finger movements of a split-brain monkey, with one eye covered, taking a food morsel from a test board that minimized tactile contrast between the food morsel and background. Under guidance of the contralateral eye (left column) the index finger and thumb dislodge the food morsel from the well. Under guidance of the ipsilateral eye (right column) the hand is brought to the proper place but the food morsel is not taken from the well. Instead, hand and fingers explore the board's surface as if blind. Redrawn from films. (From Brinkman and Kuypers 1972, with permission.)

occipital cortex and motor centers in the frontal lobe. Although this interpretation may be basically correct, one has to consider the fact that the parieto-occipital leucotomy performed in these experiments involved posterior parietal cortical areas which are known to be critical for visuomotor control. Monkeys with posterior parietal lesions alone also make large errors in reaching for food with their arm contralateral to the lesion and are unable to correctly shape their fingers during prehension (see Chapter 6).

It remains that all these experiments concord in showing a clear dissociation between visuomotor mechanisms related to directing the hand at a target [the 'localization function' of Paillard and Beaubaton (1974)] and those related to posturing the hand in anticipation to prehension [the 'identification function', Paillard and Beaubaton (1974); see also Gazzaniga (1970)].

2.3.2. Attempts at dissociating proximal and distal motor systems in subjects with an intact corpus callosum

Several authors have attempted, in normal subjects, to find behavioural correlates of the anatomical and functional duality of the motor system as demonstrated by pathological or experimental lesions. Di Stefano *et al.* (1980) measured reaction times to lateralized visual stimuli by using movements corresponding to different effector levels. Key-pressing responses requiring the use of one finger only were used for testing the most distal level; lever-pulling responses involving an action of the elbow and the shoulder were used as a test for more proximal levels. Visual stimuli were presented in the visual field ipsilateral or contralateral to the responding segment. The results of this experiment were disappointing in that reaction times for key-pressing and lever-pulling responses were found to be almost equal. Reaction time was only 2 ms longer when the stimulus was presented contralaterally to the responding limb than when it was presented ipsilaterally. The similarity between ipsilateral (intra-hemispheric ?) and contralateral (inter-hemispheric ?) conditions was attributed by the authors to a consistent initiation of the response, whether proximal or distal, by the hemisphere contralateral to the responding limb. The longer duration taken for initiating the response in the contralateral condition was attributed to the callosal delay.

In addition, Di Stefano *et al.* (1980) failed to show a hemispheric bias in initiating the responses with either the arm or the fingers. If initiation of such simple and discrete movements were controlled preferentially by one hemisphere, one would have expected to observe a lateral bias at least for movements involving the distal segments, which are exclusively represented unilaterally. In fact, hemispheric specialization for the control of movements seems to affect more the sequential aspects of motor performance than initiation of movements. Lomas and Kimura (1976) showed that in right-handed people sequential movements (like tapping the keys of a keyboard in rapid succession) were executed more accurately and more rapidly with the right than with the left hand. The same was true whether tapping involved individual finger movements or the whole arm. In addition, interference with the motor task produced by speaking during the performance of these sequential movements decreased tapping speed only for the right hand, but indistinctively for finger or arm tapping. Lomas and Kimura concluded that motor sequences were controlled by the left hemisphere irrespective of the muscle groups involved in the movements and that the mechanisms for this control partly overlapped with those responsible for speech production. More recently, however, Todor *et al.* (1982) showed that tapping rate, although it was globally faster and less variable with the right upper limb, differed markedly according to the segment used. Tapping rate was slower with the finger alone than with the wrist or the whole arm.

Other investigators have examined the pattern and the timing of movements *directed* at targets appearing in the visual field and not only *in response* to these targets. They consistently showed that reaching movements crossing the midline, that is, directed at targets in the visual hemifield opposite to the arm, were slower and less accurate than movements directed at targets on the same side as the arm (Van der Staak 1974; Prablanc *et al.* 1979a; Fisk and Goodale 1985). The Fisk and Goodale (1985) study clearly showed that the ipsilaterally directed (uncrossed) reaches were performed very differently from the contralaterally directed (crossed) reaches. Ipsilateral reaches were initiated with a shorter latency, attained a higher maximum velocity, and were completed more quickly than contralateral ones (Fig. 2.16). The latency for a contralateral reach could be as much as 25 ms longer than for an ipsilateral reach. The difference in duration could amount to 80 ms. Maximum velocity in contralateral reaches was up to 20 per cent lower than in ipsilateral ones. In addition, the accuracy of contralateral reaches (as measured by the constant and variable errors) was significantly poorer than for the ipsilateral reaches.

Ipsilateral reaches require intra-hemispheric visuomotor processing, that is, the retinal co-ordinates of the target are processed within the same hemisphere which prepares the movement. By contrast, contralateral reaches may require transfer of target information between the two hemispheres. Shifting from intra-hemispheric to inter-hemispheric processing, however, should not represent such a difference in latency. According to the above-mentioned results of Di Stefano *et al.* (1980), the difference in latency for key-pressing responses to ipsilateral and contralateral targets (*c*. 2 ms) is not in the same order of magnitude as the difference in latency between ipsilateral and contralateral reaches (*c*. 25 ms). In addition, it is hardly conceivable that inter-hemispheric transfer of target information could affect duration and velocity of the movements.

In fact, another experiment by Fisk and Goodale (1985) clearly showed that the differences in latency, kinematics and accuracy between the two types of movements was not related to the incompatibility between retinal position of the target and the side of the arm used for the response. In this

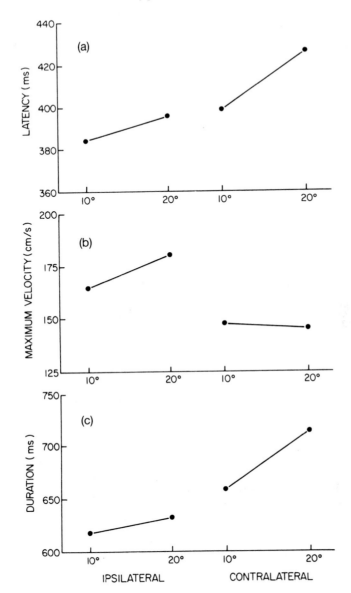

Fig. 2.16. Effect of target eccentricity on movement latency, maximum velocity and duration of ipsilateral and contralateral reaches. (a) Illustrates that while latency increased with increased target eccentricity for both ipsilateral and contra- lateral reaches, this increase was larger for contralateral reaches. (b and c) Illus- trate that for ipsilateral reaches the maximum velocity of the reach increased with increased target eccentricity, but that for contralateral reaches the duration of movement increased with increased target eccentricity. (From Fisk and Goodale 1985, with permission.)

experiment, the subjects were required to maintain visual fixation either on a point located at the centre of the screen or on other points located eccentrically with respect to body axis, during presentation of the targets. This method allowed dissociating the visual hemifield in which the target was presented, from the body hemispace in which the target-directed movement was effected. Namely, targets could be presented within the hemifield contralateral to the arm, but within the hemispace ipsilateral to that arm (Fig. 2.17). The movement kinematics were found to depend upon the spatial position of the target relative to the body axis and not upon the side of the visual field in which the target was presented. Fisk and Goodale (1985) suggested that the velocity characteristics of the reaching movements were more closely related 'to processing differences in neural systems associated with motor output' than to those related to the specification of the retinal position of the target.

An alternative explanation for the differences in the two types of reaches is that in the case of ipsilateral reaches, the target activated crossed motor pathways, although in contralateral reaches uncrossed motor pathways were involved. In other words, according to this hypothesis visuomotor processing would remain intra-hemispheric in both cases, but would involve different motor systems. There are arguments in the literature indicating that visually goal-directed movements controlled by the uncrossed pathways have poorer kinematics, and are less accurate, than movements controlled by the crossed pathway (see for instance, Brinkman and Kuypers 1973).

Looking for possible functional similarities between proximal and distal movements on the one hand, and crossed and uncrossed movements on the other hand remains highly questionable. It might be suggested that mechanisms for the command of adductive muscles (those involved in crossed movements) are less cortically organized than those for abductive muscles. Adductive movements may be considered as more body-oriented and more involved in bimanual activities, and therefore as more 'proximal' than abductive movements, which are more involved in activities oriented toward extrapersonal space. The only merit of this speculation is to suggest a tentative basis for a similar neurological organization of types of movement (proximal and crossed, distal and uncrossed) which, at first sight, have little in common.

2.3.3. The case of bimanual movements

It has long been noticed that during execution of bimanual movements, the two hands tend to become synchronized. This point was confirmed by Di Stefano *et al.* (1980) by using the key-pressing and lever-pulling responses to lateralized visual stimuli already mentioned. When subjects used both hands to produce a response, the slight advantage in favour of the limb

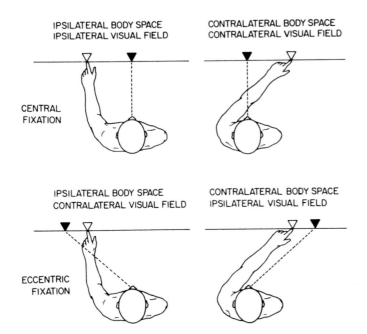

Effect of target position relative to the body axis on the kinematic measures of performance.

| | Laterality* | | Eccentricity* | |
	Ipsilateral	Contralateral	10°	20°
Maximum velocity (cm/s)	198.9	173.1	180.5	191.4
Mean velocity (cm/s)	104.0	94.0	96.7	101.3
Duration (ms)	570.4	618.4	580.9	607.9

*All differences between levels of factors were significant $F(1,3) > 11.7$, $P < 0.05$.

Fig. 2.17. Effect of hemispace on movement kinematics. By changing visual fixation from central to eccentric fixation it was possible to eliminate the effect of visual field. Ipsilateral reaches in the central fixation and in the eccentric fixation conditions correspond to targets located in opposite visual fields. By contrast, these reaches are directed at the same hemispace. The values in the table below the figure correspond to reaches directed at ipsilateral or contralateral hemispaces irrespective of the side of the visual field. Note shorter and faster movements in the ipsilateral condition. (From Fisk and Goodale 1985, with permission.)

Table 2.4. Values (in ms) of simple visuomotor reaction times (RT) in patient R.S. In the unimanual condition, RTs for the right (affected) hand (RH) were longer than those for the left (normal) hand (LH) (difference significant at $P <0.001$). In the bimanual condition RTs tended to become equal with both hands (difference between RH and LH not significant). This table is due to the courtesy of Dr F. Michel.

		UNIMANUAL	BIMANUAL
	Mean	322	337
LH	SD	93	81
	Range	207–704	240–638
	Mean	392	343
RH	SD	91	84
	Range	255–638	244–643

ipsilateral to the stimulus was preserved only for key-pressing with the fingers, not for lever-pulling with the arms. This result seems to indicate that in bimanual movements the hemisphere activated by the visual stimulus might control the proximal muscles on both sides. The same would not be possible for finger movements which can only be controlled by one hemisphere.*

There are other situations where the visual stimulus is not lateralized (i.e. the two hemispheres are equally activated), but motor responses with either arm have asymmetrical reaction times due to brain lesions. One such case was observed by F. Michel (personal communication, 1986) in one patient who presented anaesthesia of the right hand as a consequence of parietal lesion in the left hemisphere (for a full description of this patient, see Jeannerod *et al.* 1984; see also Chapter 5). When simple visuomotor reaction time (key-pressing) was tested with each hand separately, the right (affected) hand performed consistently more slowly than the normal one (Table 2.4) When both hands were used simultaneously, the reaction times tended to become equal, due to reduction of the longer reaction time of the affected hand (Table 2.4). This result indicates that the mechanisms for initiating bimanual movements are not the same as those for unimanual movements. By using the same speculation as above, it might be argued that bimanual movements are more proximally organized (and are less

---

* In bimanual movements, there is asymmetric division of labour between the two hands. In right-handers, the right hand usually achieves the manipulative component while the left hand has a postural role. Indeed, the right hand is capable of a greater accuracy than the left one (e.g. Flowers 1975; for a complete review, see Hécaen 1984).

dependent on cortical control) than unimanual movements. Associated movements observed in hemiplegic patients would be justiciable of the same explanation.

These findings can be interpreted within the framework of a hierarchy of the operations occurring in a motor program. According to this interpretation, certain dimensions of a programmed movement would have to be necessarily activated prior to other dimensions of the same movement. Studies using reaction time as an index have shown that giving information about the goal of a movement shortly before its execution reduces its latency, hence indicating faster activation of the motor apparatus by the program (e.g. Rosenbaum 1980). This technique of 'pre-cueing' can be used as a tool for probing the order of activation of program components, by providing information specifically related to one or the other of these postulated components. For instance, pre-cueing the arm (right or left) to be used in a pointing task and the direction where the target will appear (e.g. upward or downward) has a more beneficial effect on the latency of the movement than pre-cueing the distance of the target (Rosenbaum 1980; for review, see Requin 1985). One can therefore postulate that the optimal order of activation would be determined by the degree of 'proximality' of the components of an action. In the above examples of bimanual movements, it could be that activating synchronously the two arms would represent a simpler operation than selecting one limb. Subsequently, other operations like determining the direction in which to move and the extent of the movement would come into play. Breaking this hierarchy might be one of the problems occurring during learning of motor skills requiring inter-limb independence, like music performance (e.g. Schaffer 1981).

3. The role of visual feedback in movement control

Most movements executed in everyday life are aimed at visual objects and are controlled visually. Part of this control is exerted by feedback signals arising from the movement itself. Controversies about the modes of action of visual feedback have generated interesting concepts (e.g. intermittent vs continuous control, speed–accuracy tradeoff vs. impulse variability) which are still essential nowadays for understanding visuomotor performance.

Visual feedback can be easily manipulated experimentally. Until recently, terminal accuracy of movements was the most commonly considered experimental variable for testing the effects of these manipulations. Accuracy, however, is the end result of the visuomotor process as a whole, and gives only limited information on the intervening mechanisms. The introduction of kinematic measurements as additional experimental variables has provided a new approach for determining the level at which visual feedback impinges upon the system to improve its performance.

Selective alteration of visual feedback also allows inferences to be made about the nature of the visual signals at work in the control of movements and the brain structures where they are monitored. Experimental data support a dissociation of mechanisms subserving control of reaching by central and peripheral visual fields, respectively.

3.1 Changing views on visual control of movements

3.1.1. Woodworth and the notion of sensory feedback

The first experimental contribution to the problem of visual control of movement was that of R.S. Woodworth in 1899. In his long monograph published in the *Psychological Review*, Woodworth established many of the modern concepts which are still valid today for understanding the determinants of accuracy of goal-directed movements. Thus, it may be useful to review in some detail the main aspects of this contribution in the light of our present knowledge of the problem.

Woodworth's data were based on experiments with normal subjects. He used a recording paper chart displaced by a kymograph rotating on a horizontal axis. The paper was visible to the subject through a slot of variable width parallel to the axis of the kymograph, i.e. perpendicular to the

84

direction of the paper displacement. The subject's task consisted of ruling lines along the edge of the slot with a hand-held pencil. The displacement of the pencil was directly recorded on the paper.

A stop was used as a starting point. The target to which the movements were directed could be either a 'normal' line which the subject was to copy, or a mark on the paper at which the movements were required to terminate. More often, however, the subject was required to draw the line equal to the one he had made just before. In that case, the width of the slot was adjusted so that the subject could see only that line. The total extent of movements was around 15 cm. A metronome which, in Woodworth's terms was his 'constant companion' throughout the work, was used to pace the movements. The stroke rate could be adjusted between 20/min and 200/min, which allowed the inter-movement interval to be varied between 3 s and 300 ms.

For processing his results, Woodworth used methods introduced by Fechner for the measurement of psychophysical thresholds, such as the 'method of right and wrong cases' of Fechner, and the 'method of average error'. He considered the latter as the most appropriate. By measuring the difference in length of the lines with respect to each other, Woodworth easily obtained the 'total error' (the sum of individual errors in each direction) and the 'average error'. The 'constant error' was also obtained as the difference in length between the first and the last line in the series, divided by the number of lines less one.

By varying systematically the rate of repetition of the movements Woodworth discovered that the average error clearly increased with the stroke rate of the metronome, reaching its maximum value at a rate of about 140–160 strokes min^{-1} (i.e. when inter-movement interval was down to about 400 ms). At higher rates, average error kept the same value and was not further influenced by decrease of intermovement interval (Fig. 3.1). A logical conclusion for this result was that the subject's opportunity to control movement accuracy was limited by the duration of the movement, itself depending on speed of execution. Further validation of the relation between speed and accuracy of movements was obtained by Woodworth in a condition where intermovement interval was kept constant while speed was varied independently, by instructing subjects to draw slowly or rapidly. The result of his experiment was that fast movements yielded to much larger error than slow movements.

The other important finding reported by Woodworth (1899) was that error in the execution of rapid movements is in fact due to lack of visual control. Indeed, when subjects were required to repeat the task of drawing lines in synchrony with metronome strokes with their eyes closed, they made similarly large errors for low as well as high stroke rates. For each subject the average error produced in executing movements at any speed with eyes closed was within the same range as that produced in executing movements rapidly with eyes open. The resulting diagram of error vs inter-

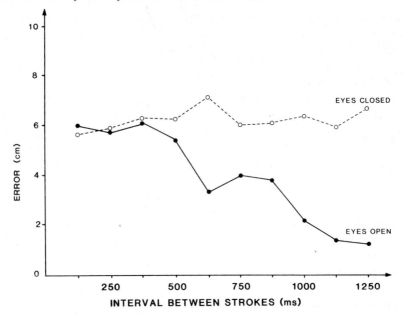

Fig. 3.1.   Relation of accuracy to speed with and without feedback on voluntary visually goal-directed movements. The data from Table 3.1 (right hand) of Wood-worth (1899) have been replotted. In this experiment, movements were paced by metronome strokes. The interval between strokes (calculated from stroke frequencies given by Woodworth) gives an idea of movement speed.

Note no difference in error amplitudes in eyes-open and eyes-closed conditions for rapid movements (short inter-stroke intervals). This, according to Woodworth, represents the 'bad effect of speed'.

movement interval was thus a horizontal line without evidence of a speed–accuracy co-variation (Fig. 3.1). In addition a constant error was observed with eyes closed, i.e. movements tended to be too long when executed at a fast speed and too short when executed at a slower speed. From these results, Woodworth inferred that the 'bad effect of speed' on movement accuracy was in fact due to the subjects' inability to control their movements visually when their duration became shorter than a critical value.

The relationship between speed and accuracy of movements and the role of visual control were interpreted by Woodworth in terms of interplay between 'initial adjustment' and 'current control'. The initial adjustment reflected, in Woodworth's terms, the *intention* of the movement as a whole, i.e. its co-ordination and its extent. An error in initial adjustment produced misdirection of the movement. On the other hand, current control was responsible for final adjustment of the movement. 'The bad effect of speed', Woodworth thought, 'consists in rendering impossible a delicate *current control* in preventing those later and finer adjustments by means of which a movement is enabled to approximate more and more closely to its goal' (Woodworth 1899, p. 42). This fact was substantiated by the observation that movements executed at normal speed with eyes open are slower at the end or even completed by small additional movements, although movements executed at high speed lack these final adjustments. Final adjustment produced by current control might be, according to Woodworth, an important factor in learning skills. After many repetitions of the same movement final adjustment would become progressively less necessary and accuracy would be obtained by improving the precision of the initial component. 'Whether the great virtuosos do away entirely with the later adjustments and achieve their wonderful accuracy by means of the first impulse, would be an interesting thing to find out' (p. 54).

Woodworth considered that sensory cues used for current control of the movement had to be sensations coming from the moving limb itself. He thought that normally, visual and kinaesthetic sensations collaborate for movement control, but that visual cues cannot substitute for kinaesthetic cues, 'simply because we never have looked at the movements of our fingers . . . but always at the result we are accomplishing. Consequently we have no association between the visual sensation of the moving fingers and the proper impulse to set the muscles into coordinated action.' (p. 76).

Woodworth's monograph is not only interesting from a historical point of view; it also contains a number of experimental results which are still largely valid. The methodology it describes is still considered as the basic paradigm for study of movement control. When Woodworth's experiments were replicated about 60 years later by Vince (1948), his results were largely confirmed. The only substantial improvement introduced by the latter author was a recording of the movement on a faster-rotating drum, thus allowing precise measurement of movement duration. By this method Vince was able to determine that an adverse effect of movement speed on accuracy was already present when movement duration was less than 1400 ms. For movement durations below 750 ms, accuracy declined further (average error amounting to 2–4 per cent of movement amplitude). Finally below 500 ms, error amounted to 7.5 per cent of movement amplitude. The interpretation of these results by Vince was that visual feedback was excluded when movements lasted less than 750 ms, and that kinaesthetic feedback was also excluded in movements lasting less than 500 ms.

Another important contribution of Woodworth was his remarkable intuitions concerning mechanisms of movement control. One of these intuitions was that 'it is the accuracy of a movement that makes it careful and purposive. While some few movements require only brute force of a comparatively ungoverned sort, in most cases there must be a considerable degree of control and adaptation to a particular end.' (p. 4). The distinction made by Woodworth between ungoverned and controlled movements or, within movement itself, between initial impulse and current control already includes an implicit conception of 'feedback regulation'. This concept, which seems to have been introduced in biology by Pflüger, originally stated that actions were governed by a teleological state or *goal* and that the mechanisms which produced these actions were regulated by the difference between the goal and the degree of execution (see Henn 1971). The fundamental notion that *regulation implies a goal* was familiar to Woodworth when he designed his experiments on 'automatic' movements. These aimless movements differed from movements directed at targets in that they lacked regulation, i.e. they had no corrections or slowing down at the end of the trajectory.

Woodworth's conception of movement control was that of a process operating continuously until the goal was achieved. This is the classical design of a continuous feedback with a pure delay (needed for transmission of control inputs to the motor system). This design, according to Woodworth, accounted for the fact that fast movements (i.e. shorter than the delay) appeared to be virtually uncontrolled, and that in movements of longer duration control could be exerted only after the delay had elapsed, that is near the end of the trajectory.

3.1.2. The intermittent feedback control theory

A further step in understanding movement control was made during the Second World War, and was conceptualized in an influential paper by K. Craik, published in 1947. Craik's contribution must be understood within the context of development of information theory and its application to machine control by the human operator. One major problem met during this period by engineers who built devices that could track moving targets, was to assess the limitations of the man–machine system. Human movement was thus not the object of study in itself, it was embedded in the man–machine operation. This may have favoured the use of tracking tasks for testing the performance of the human operator. In such tasks (e.g. moving a handle to track a slowly moving spot on a screen) subjects appear to make intermittent corrections spaced by about half a second. In other words, according to Craik, 'the human operator behaves basically as an intermittent servo'. The simplest explanation for the 0.5 s intermittency period was the possible existence of a dead zone needed for feedback signals to reach a level of intensity sufficient to be perceived by the control system. In fact, Craik considered this explanation as invalid, mainly because corrections generated by the human operator were of a ballistic type. He used the term ballistic to indicate that the correction was triggered off as a whole, not on a continuous basis. Ballistic behaviour contrasts with that of an intermittent servo, where the correction would run until error noticed during the sampling period is nullified. Craik argued that continuous correction in a system with such a long response time would inevitably generate oscillations at the end of the movement, since correction would continue during the last sampling period and finally exceed the error.

A more probable explanation for the behaviour of the human operator would be that of a system periodically generating ballistic corrections, with a constant undershooting of target position. During tracking the system would consistently lag target position; when the target stops the movement would approximate its position after a delay corresponding to one sampling period. Assuming, as Craik did, that ballistic corrections are too short by about 10 per cent with respect to error, the final approximation after the target stop would drive the effector to within 1 per cent of target position. In addition, such a system would be devoid of oscillations around a stationary target, due to the insensitivity of ballistic systems to feedback signals during the movement itself. A further advantage of discontinuous feedback, according to Craik, would be the maintenance of a high precision in rapid movements where continuous feedback is necessarily excluded. In rapid movements, 'sensory feedback must take the form of a delayed modification of the amplitude of subsequent movements. Sensory control, in other words, alters the amplification of the operator with a time-lag and determines whether subsequent corrective movements will be made; it does not govern the amplitude of each individual movement while it is being made.' (Craik 1947, p. 58).

The nature and duration of the time-lag responsible for discontinuity of visual feedback in movement control was a major concern for Craik and his collaborator M.A. Vince. The typical 500 ms interval observed between successive corrections in visual tracking tasks was thought to correspond to the phychological refractory period, such that a response 'is organized and begin, before the organizing centre is impinged upon by a second stimulus' (Vince 1948, p. 86). According to Vince this 500 ms time had to be subdivided into a visual reaction time (*c.* 250 ms) and a motor time (another 250 ms). This time constraint would thus determine the *maximum rate* at which a visual stimulus could be used for controlling a movement.

3.1.3. Fitts' law

Later, in the context of the information theory, the basic intermittency observed in the human operator during continuous tasks was attributed to the single-channel nature of the central mechanisms which detected the signal and generated the responses. According to this interpretation one signal had to be cleared before another one could be dealt with. Intermittency merely reflected 'queueing' of the stimuli which successively occupy the channel (Davis 1957). Errors appearing in movement execution were thus explained by the limited information capacity of the sensorimotor channel controlling the movement in response to a stimulus. This idea was first put forward by P. Fitts in his paper published in 1954. Fitts described the information capacity of the motor system as 'specified by its ability to produce

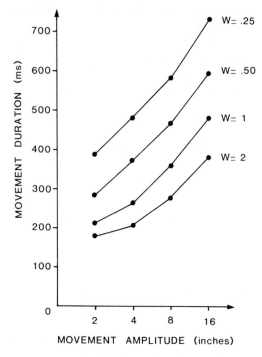

Fig. 3.2.   Relationship of movement duration to movement amplitude in the Fitts reciprocal tapping task. Data from Table 1 of Fitts (1954) have been plotted as a function of movement amplitude (in inches) and target width (*W*, in inches). Note strong movement duration to amplitude relationship for each target width. Also, note global increase in movement duration when target width decreases.

consistently one class of movement from among several alternative movement classes. The greater the number of alternative classes, the greater is the information capacity of a particular type of response.' (Fitts 1954, p. 381). If movement parameters (like force, direction, and amplitude) were controlled as experimental variables, information capacity of the responses was only limited by their statistical variability or 'noise'. Therefore the information capacity could be inferred from measurements of statistical variability.

In one of the best known Fitts' experiments [the reciprocal tapping task; Fitts (1954)] subjects had to hit alternately with a stylus two targets of variable width (between 0.25 and 2 in.), separated by a variable distance (between 2 and 16 in.). An error was scored whenever the stylus undershot or overshot target position by any amount (the 'hit or miss' score). In principle, instructions given to subjects emphasized both accuracy and speed of the movements. In fact, accuracy seems to have been more reinforced than speed in the experiments by Fitts and his followers.

The main finding was that for each given target width, movement duration increased with movement amplitude (Fig. 3.2). In addition, for a given distance between targets, movement duration increased when target width was decreased. The proportion of errors also increased when the task was made more difficult i.e., when either target width was decreased, or target distance increased, or both. However, in the reciprocal tapping task, the error rate never exceeded about 4 per cent, which justified data analysis involving all the movements, including those where a miss was recorded.

Fitts concluded from his experiments that speed, amplitude, and accuracy of a movement made under visual control were linked by a three-way relationship. He considered that a motor task could be defined by its *index of difficulty*:

I_d: $\log_2 2A/W$ (in bits per response)

where W is the target width and A the distance between targets (movement amplitude).

The corresponding movement was characterized by its *index of performance*:

$I_p = 1/T \log_2 2A/W$ (in bits per second)

where T is the movement duration.

The above three-way relationship thus can be condensed into a single equation linking movement duration to the index of difficulty. Later, this equation became known as 'Fitts' law' (Crossman and Goodeve 1963):

$T = a + b \log_2 2A/W$

where a and b are empirically fitted coefficients.

As shown in Fig. 3.3, replotted from Fitts' data, the empirical function relating movement duration to index of difficulty is fairly linear over a wide range of conditions. According to Fitts, this relation reflected the subject's trade-off between velocity and accuracy*. This assumption implies that when accuracy of a movement had to be preserved, the time taken for this movement was determined by the amount of information which had to be transmitted in accomplishing the movement. Assuming that the transmission channel between input and output is 'noisy', the greater the index of difficulty, the longer it takes to extract the signal from the noise.

Fitts' law represents an adequate description of the covariation of spatio-temporal parameters of goal-directed movements. It sets the limits within which these parameters can be varied, and control by sensory feedback exerted. But the explanation given by Fitts to these limits, based on the information capacity of the sensorimotor channel, has been questioned. An alternative explanation was proposed by Crossman and Goodeve in 1963 (see Crossman and Goodeve 1983). Their hypothesis was that

---

* It is interesting to note that the logarithmic shape of the function described by Fitts for the speed/accuracy trade-off seems to hold mainly if accuracy of the movements is reinforced (spatially constrained movements) although it takes the form of a linear function if it is the speed of the movements which is reinforced (temporally constrained movements) (e.g. Meyer *et al.* 1982).

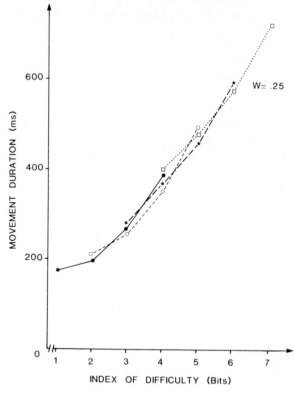

Fig. 3.3.   Relationship of movement duration to index of difficulty (Fitts' law). The same data as in Fig. 3.2 have been replotted. Index of difficulty (in bits) is calculated as $I_d = \log_2 2A/W$ (A = amplitude of movements, W = width of targets). Note quasi-linear relationship for I_d above 1. Different lines and symbols represent different target widths. Dark circles: $W = 2$ in. Open circles: $W = 1$ in. Stars: $W = 0.5$ in. Squares: $W = 0.25$ in.

'if limb position is adjusted by making limb velocity negatively proportional to error [by way of a velocity feedback] the limb's response to a step function command input (. . .) will be an exponential approach to the new target with time constant dependent on loop gain. If the motion terminated at the edge of a tolerance band, this predicts a total time proportional to the logarithm of motion amplitude divided by width of tolerance band' (Crossman and Goodeve 1983, p. 256)

In order to demonstrate their hypothesis, the authors replicated the Fitts reciprocal tapping experiment, with a method enabling them to record the hand trajectory. Records were obtained by monitoring the slide of a thread attached to the stylus. Movements were executed in two different conditions. In condition A movements had an endpoint constrained by small targets (e.g. 0.25 in.), as in Fitts' experiment. In condition B the targets were so large (e.g. 16 in.) that movements were virtually 'aimless'. Crossman and Goodeve found that the deceleration phase of the velocity profile of movements executed in condition A was a good approximation of an exponential function. By contrast, in condition B the velocity profile was symmetrical, i.e. lacked the exponential decay in velocity before the stop. The authors considered that the symmetrical velocity curve observed in aimless movements was the result of a single command 'impulse', a hypothesis already put forward by Woodworth. In addition Crossman and Goodeve explained the exponential velocity curve observed in precise movements by a succession of overlapping and diminishing smaller curves of symmetrical shape. In other words, they thought that precise movements were in fact composed of iterative single pulses approximating more and more closely to the target position. In some of their records, Crossman and Goodeve were able to observe directly a ripple, indicating the presence of individually distinguishable impulses.

This hypothesis is an interesting one, because it adapts Craik's model (designed to account for continuous tracking) to discontinuous tracking or step responses. According to Crossman and Goodeve, 'the physiological mechanisms needed to implement this type of feedback process would be quite simple, producing intermittent repositioning impulses of stereotyped form whose amplitude could be modulated according to current error as estimated at the time of initiation . . . ' (Crossman and Goodeve 1983, p. 267). In order to comply with Fitts' estimate of an information rate of 10 bits s^{-1} for the sensorimotor channel, the authors had postulated for their feedback model a sampling rate of about 10 per second. Such a high value raised the problem of how a complete response to the feedback signal (from error estimation to corrective impulse) could take place within one sampling period, considering the relatively long duration of kinaesthetic and visual feedback times (see below). One possible explanation proposed by Crossman and Goodeve to account for this difficulty was that the error would be estimated, not from limb position, but from the driving signal to the muscle. This hypothesis, however, seriously departs from the usual concept of feedback mechanisms.

A similar feedback interpretation of Fitts' law was also developed later by Keele (1968, 1981). According to this author duration of the initial movement, executed before correction by visual feedback can be made, should be independent of distance traversed, the movement time to distance relationship implied in Fitts' law being produced by the corrective submovements executed after visual feedback. Similarly, it should also be independent of final precision, since precision would be achieved by the corrections. Indeed, Annett *et al.* (1958) had found that, in prehension movements, the initial fast distance-covering phase was independent of final precision. Empirical evidence for the visual feedback interpretation of Fitts' law can be found in several experiments. Klapp (1975) for instance, using a reciprocal tapping task in which distance between targets could be varied between 2 mm and 33 cm, showed that Fitts' law held only for movements lasting more than 200 ms. In the case of 'easy' movements (i.e. small movements directed at large targets) lasting less than 200 ms, no relation was found between the index of difficulty and movement duration. Klapp suggested that short-lasting movements had to be preprogrammed and were not subjected to feedback control. This suggestion was based on results from a different experiment (Klapp 1975), where visual feedback information was suppressed at the onset of the movement. This resulted in a dramatic increase in error rate for large (difficult) movements (error rate jumped from 4.4 to 93 per cent) but not for small (easy) movements (error increased from 1.6 to 10 per cent). Thus large movements are likely to be composed of one programmed component followed by one or several feed-back-dependent components, although small movements would involve only the programmed component, with little or no influence from visual feedback. Consequently the small movements deviate from Fitts' law.

The exception to Fitts' law provided by small movements, as postulated by Klapp (1975), was not confirmed by Langolf *et al.* (1978). In their exper-iment these authors also used the reciprocal tapping task, with the require-ment that errors should be limited to 5 per cent of trials (hit and miss score). Two ranges of movement amplitude were used: a microscopic range (0.25–1.27 cm) and a range similar to that of Fitts (1954), with ampli-tudes up to 30.5 cm. In both cases movements times appeared to be related linearly to the index of difficulty of the task. Movement recordings showed that the velocity curves for microscopic movements were discontinuous, thus indicating existence of at least two distinct movements. For move-ments of 1.27 cm amplitude for example, the trajectory was composed of a primary movement lasting on average 195 ms and bringing the hand an average distance of about 93 per cent of the total target distance. The second movement had approximately the same duration as the initial one and corrected for the remaining 7 per cent of target distance. For move-ments of a smaller amplitude, the initial movement lasted 160 ms and cor-responded to about 90 per cent of the trajectory. Again, duration of the second movement was the same as that of the initial movement. It thus seems that, for small movements (i.e. effected mostly by the fingers) the linear interactive model postulated by Crossman and Goodeve (1963) and by Keele (1968) to explain Fitts' law would be in good correspondence with the data. This was not so clear with movements of larger amplitudes (wrist or elbow movements). In that case, movements trajectories tended to have a smooth deceleration without evidence of corrective movements. Lack of visible secondary movements, however, might have been explained by a higher inertial damping of the upper segments of the arm as opposed to the fingers. Nevertheless, the shape of these trajectories changed when the tolerance of the movement (1/W) was changed, i.e. the time for the initial movement was decreased when the target size was smaller. Since Fitts' law still held for these movements, this change could not be explained by the constant duration, constant accuracy discrete corrections implied in a linear feedback model.

More recently, the validity of this model in accounting for Fitts' law was re-examined by Jagacinski *et al.* (1980). In their experiment, subjects had to capture a target appearing on a screen by rapidly moving a joystick. Target direction (right or left), distance (48–147 mm), and width (3–9.2 mm) were manipulated. The speed of the movements was emphasized. Subjects had to hold on target position by using either position or velocity feedback. As Fitts' law would predict, movement time appeared to be linearly related to the index of difficulty. The microstructure of movements was determined by recording joystick position. The curves showed a large amplitude initial movement followed by other submovements of a smaller amplitude. Accuracy of the initial movement was clearly a function of its duration, in the sense that it covered a greater distance to the target centre. This result disproved the simple iterative model for Fitts' law since the successive submovements were found to have different durations and different amplitudes.

To account for this and other discrepancies, another interpretation of Fitts' law has been proposed, in which it is assumed that the relationship of movement time to index of difficulty is already present within the initial impulse, i.e. at the programming stage of the movement. As set out by Meyer *et al.* (1982), this interpretation assumes that in target-aimed movements, the initial impulse is programmed to hit the target and that errors which are observed in this type of movement reflect neural variability of initial impulses. According to their impulse model, variability is thought to increase proportionally with movement velocity. Consequently, duration of the initial impulse should be optimized in order to keep movement velocity within a range compatible with the accuracy demand. According to Abrams *et al.* (1983), such a model can be demonstrated mathematically to closely approximate the logarithmic speed–accuracy trade-off. Among the predictions of the optimized impulse model is that a similar speed–accuracy trade-off should be present also in movements executed in absence of visual feedback. Results partly confirming this prediction can be found in the literature (Prablanc *et al.* 1979a; Wallace and Newell 1983), to which Abrams *et al.* (1983) have added new evidence.

3.2. Time for visual feedback to influence movements

Precise determination of the time needed for a visual input signal to influence an ongoing movement is a central issue for establishing the role of visual feedback in motor control. Early investigators considered visual feedback as relatively slow. According to Woodworth (1899) repetitive movements separated by 1 s or less tended to become more and more inaccurate. When this interval was reduced to about 400 ms, he assumed that no visual control could be exerted on the movement. Vince (1948) made similarly conservative estimates, since she thought that correction in the course of a movement could not occur more frequently than once every 500 ms.

3.2.1. The Keele and Posner experiment

The first attempt at measuring directly visual feedback time was made by Keele and Posner (1968). These authors used an experimental design where subjects made discrete pointing movements toward small targets located 6 in. (14.4 cm) away from starting position. Movement time was measured to the nearest 10 ms between lifting the stylus from the starting position and contact at or near the target. The signal to move was given by illumination of the target, which was turned off as soon as the stylus was lifted. In addition a general illumination of the display was provided by a light which could also be turned off at a movement onset. Half of the trials were made with the light on and half with the light off (i.e. in complete darkness). Trials were grouped in sequences where subjects attempted to pace the speed of their movements so that movement duration would approximate 150 ms, 250 ms, 350 ms, or 450 ms, respectively. Results were scored, as in the Fitts' experiment, as the proportion of 'misses' with respect to the total number of movements in each block of trials. Precise amplitude of errors was not measured.

   The graph in Fig. 3.4 shows the proportion of misses against mean intended movement time for movements executed with visual feedback (i.e. with lights on) and without feedback (lights off). Three main facts seem to arise from this experiment. First, the proportion of errors is significantly higher in the 'no visual feedback' condition when movements last for more than about 200 ms. The longer the movement duration the greater the difference. The second fact is a global decrease in rate of errors when movement time increases, in both visual-feedback and no-visual-feedback conditions. Keele and Posner attributed this effect to the role of kinaesthetic feedback. Finally, movements executed in the 'no visual feedback' condition were of a shorter duration (by about 20 ms) than the corresponding movements executed with vision. This was true for each class of intended duration, except for the first class (150 ms) (Fig. 3.4). This latter result, not picked out by the authors, will be discussed later in the context of the study of velocity pattern of movements executed with or without visual feedback. Keele and Posner concluded from their findings that it takes between 190 and 260 ms for visual feedback to influence accuracy of an ongoing movement.

   A closely similar value was obtained by Beggs and Howarth (1970) in an experiment where accuracy was measured as a function of movement

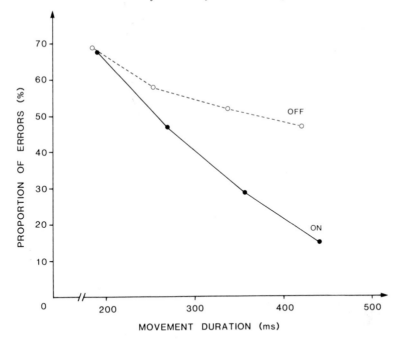

Fig. 3.4. Proportion of errors as a function of movement duration in a pointing task. Data of Table 1 of Keele and Posner (1968) have been plotted as a function of movement duration. On: pointing movements executed under visual control. Off: Pointing movements executed in the dark. Subjects attempted to produce movements of 150, 250, 350, or 450 ms duration.

In the 'on' condition, there was a relatively good correspondence between intended and actual movement durations. By contrast, in the 'off' condition, movements were shorter. Note increase in error rate in both conditions for rapid movements. Compare this result with that of Woodworth (1899) shown in Fig. 3.1.

duration and time at which the visual feedback loop was interrupted. The subjects task consisted in pointing a pencil at a target in front of their face starting from a home position near their shoulder. Alternating, metronome-paced movements were performed. The assumption was that 'the accuracy of hitting the target should depend on whether the lights were turned off nearer or farther from the target than a distance corresponding to the corrective reaction time . . . ' (p. 752). The accuracy should be greater when this time is exceeded. Analysis of the results showed that errors increased when the lights were turned off less than 290 ms before the end of a movement performed at any of the tested speeds.

Following these experiments, a time of 200–300 ms for processing visual feedback and correcting movements became accepted by most authors. This value was considered satisfying because it closely corresponded to the typical visual reaction time recorded for hand movements, in response to unexpectedly appearing visual targets. An average value of approximately 250 ms for the latency of hand responses has been found empirically by Stark (1968). This author discussed the problem of whether the sampler which determines the intermittency of responses should be synchronized to the appearance of the stimulus or to an independent clock. He found that varying the gain of the visual feedback loop during the hand movement (by feeding the hand position signal into target position) produced a succession of regularly spaced hand 'saccades' while the subject attempted to track target position. These saccades occurred at a rate of 4–5 per second, a value which would account for the average value of 250 ms found for the latency of single-step movements.

There are indications, however, that visual feedback may be processed and used in less time. These indications can be drawn from more recent replications of the Keele and Posner experiment. Such a replication was attempted by Zelaznik *et al.* (1983), who performed a series of experiments with the aim of sorting out the main questions raised by the original results of Keele and Posner. In the experiments by Zelaznick *et al.*, subjects had to displace a hand-held stylus to a small target (1 mm) located at a distance of 10 cm (index of difficulty was around 1.01, to be compared to that of 1.04 in Keele and Posner's experiment). Movement duration was manipulated between 100 ms and 240 ms by instructions given to subjects, and was measured to the nearest 1 ms. A microswitch, released by onset of movement, activated a circuit which determined whether lights would stay on throughout the movement or would be turned off at movement onset. The main concern of Zelaznick *et al.* was that uncertainty about presence or absence of visual feedback when a movement is started could delay the use of visuomotor cues and thus lengthen visual feedback processing times. Keele and Posner had used a situation where the probability for presence of visual feedback during the movement was $P = 0.5$; Zelaznick *et al.* compared the same situation with another one where visual feedback was constantly available ($P=1$). They found that in either condition ($P = 0.5$ or $P = 1$) presence of feedback had a positive effect on accuracy for movements of a duration as short as 120 ms. This result seriously departs from that of Keele and Posner. Even though procedures in scoring the results were not identical in the two experiments (Keele and Posner used a hit and miss score, whereas Zelaznick *et al.* used a measurement of real errors in mm) this cannot account for such a large difference in estimation of visual feedback processing time. In addition, Zelaznick *et al.* reported another experiment where accuracy was measured for movements of very short duration (70 ms). They found no effect of presence or absence of visual feedback on errors in distance, although errors in direction appeared to be reduced when visual feedback was present. The results of Zelaznick *et al.* were replicated recently by Elliott and Allard (1985). According to the latter authors, subjects tended to benefit more from visual information (i.e. directional error was reduced even for very short movement durations) when they were certain that it would be available during the movement. On the contrary when subjects were uncertain as to whether or not feedback would be available, they behaved 'like Keele and Posner's subjects', that is, they did not use visual feedback and performed fast movements in a visual open-loop fashion. In other words, time was not the only limiting factor for the use of visual feedback in movement control.*

Other factors like the type of strategy induced by the experimental conditions might also be critical. Following the same idea Elliott and Allard (1985) performed other experiments where visual feedback, when present, was distorted by prismatic goggles. The power of the prism was changed randomly across trials so that no adaptation could occur. This procedure was used to force subjects to rely on visual feedback to achieve accuracy. Again, a positive effect on accuracy was found in the condition with (distorted) visual feedback. This result suggests that visual feedback error signals concerning movement direction can be used very early after a movement has been started.

It should be mentioned, however, that another replication of the Keele and Posner type of experiment, made by Wallace and Newell (1983) yielded somewhat different results. Wallace and Newell used a Fitts type of paradigm where subjects aimed at targets of variable width (0.64–2.54 cm) placed at a variable distance (3.8–30.1 cm in the frontal plane) from the starting position. Movement duration and error rate (hit and miss score) were measured. The relation between movement duration and index of difficulty also held when the same movements were executed in the dark. Exclusion of visual feedback affected accuracy when the target distance–size combination corresponded to an index of difficulty greater than 3.58. In such conditions, movement time was generally longer than 200 ms. This result seems to exclude the idea that movements lasting less than 200 ms can be modified by visual feedback.

3.2.2. Experiments with selective exclusion of visual feedback

Selective exclusion of visual feedback during part of a movement also provides evidence for a visual feedback processing time shorter than 200 ms. In an experiment published in 1976, Conti and Beaubaton had subjects pointing at targets located 20 cm from the starting position. By

---

* There is an asymmetry between the two hands for what concerns accuracy and duration of aiming movements. In right-handers, the right hand exhibits a lower error rate and its movements are faster than those of the left hand (Annett *et al.* 1979). Movements with the left hand present more frequent changes in acceleration, suggesting more error corrections (Todor and Cisneros 1985). It has been argued that these differences could be due to a better ability of the right-hand control system to use visual feedback (Flowers 1975; Roy 1983). In fact, Roy and Elliott (1986) showed that suppression of visual feedback affected the speed–accuracy trade-off for both hands to the same extent.

instruction, movements were intended to be either 'fast' (less than 200 ms), 'normal' (between 200 ms and 700 ms) or 'slow' (longer than 700 ms). Vision of the hand could be allowed or prevented, or limited to the early, intermediate, or final parts of the trajectory. In the condition where visual feedback was not available, errors had the same amplitude for all three classes of movement. Movements with normal visual feedback, or visual feedback limited to the final part of the trajectory, were accurate provided they were executed at a 'normal' or 'slow' speed. This result is basically a confirmation of the results of Beggs and Howarth (1970). However the large variability of movement duration within each class does not allow a firm conclusion to be drawn as to the minimum duration required by terminal feedback to influence movement accuracy. Another aspect of Conti and Beaubaton's results was that movements with visual feedback limited to the early part of the hand trajectory were more accurate than movements executed without any visual feedback. Presumably visuomotor cues related to the early phase of the movement could be stored and used at a later stage of the same movement to improve its precision. This point will be further emphasized in another section (see Chapter 5).

A systematic study by Carlton (1981) also favours relatively short visual feedback processing times. In his first experiment subjects had to move a stylus as rapidly as possible toward a target (target width $W = 1.27$ cm) located 32 or 64 cm from the starting position. Movement duration and error rate (hit and miss score) were recorded. A movable shield could be used to mask a portion of the initial distance of the movement. Trials were made with 25, 50, 75, or 93 per cent of movement trajectory masked, and compared with trials where full vision of the movement was available. The results showed that error rate increased when more than 50 per cent of the hand trajectory was masked, hence indicating that visual cues related to the initial portion of the trajectory would have little influence on movement accuracy, in contradiction with other studies.

In order to make a precise estimate of the time needed to improve accuracy during the second half of a movement, Carlton designed a second experiment where he recorded movement trajectories by high-speed cine-film. Only the target located at 32 cm was used, and the shield was used in the positions where it masked 75 or 93 per cent of the initial movement distance. Position, velocity, and acceleration profiles of the stylus displacement were reconstructed from film, and the time between the stylus becoming visible and the beginning of corrective movement was measured. An average time of about 135 ms was found in three subjects. Masking a longer portion of movement distance (e.g. 93 per cent) resulted in delaying the corrective movement and lengthening total movement duration. This experiment confirms first that visual feedback is more important during the late part of the movement, and second that visual feedback time may be as short as 135 ms.

3.2.3. Time to react to a moving visual stimulus

Studies of hand movements directed at moving visual targets have also provided some evidence (though much more controversial) that relatively short times are sufficient for processing visual feedback. Whiting *et al.* (1970) designed a task where subjects had to catch by hand a ball thrown in their direction. The ball was illuminated from inside by a bulb which could be turned off 'in flight'. Several conditions were used where the ball could be illuminated for durations ranging between 100 ms and 400 ms or more. The results showed that the subjects' performance (i.e. proportion of catches) increased with the duration of illumination. About half the subjects made good catches with durations of illumination as short as 100 or 150 ms. In this experiment, however, the ball was illuminated from the beginning of its trajectory and light was turned off at a prescribed time after launching. In other words, even though the illumination period could be as short as 100 ms, the catch occurred after a relatively constant duration (the duration of the ball flight, *c*.400 ms). Therefore, the Whiting *et al.* result indicates only that 100 ms is enough time for sampling visual information and for predicting where the ball will be at the end of its trajectory. It is interesting to note, however, that the catching rate continued to improve (up to 100 per cent of catches) right up to keeping the light on for the total duration of the ball flight. This result indicates that the last 100 ms of the ball's flight provided visual feedback which could be used for adjusting hand position in anticipation of the catch.

A similar conclusion was reached by Lee *et al.* (1983) in their ball-punching experiment. These authors analysed knee and elbow angles as a function of time during the act of leaping to punch a falling ball. They hypothesized that subjects used optic information to specify the time remaining before contact with the ball. In the case of a falling ball, time to contact can be specified by a visual parameter (τ), which includes both instantaneous distance and velocity or acceleration of the ball. If joint angles are geared to the parameter τ they must be a function of a value of τ extrapolated from the latest known value of this parameter. In other words, joint angles must be geared to a value of τ corresponding to time to contact minus visuomotor delay. With this method, Lee *et al.* showed that rapid changes in joint angles occurred some 50 to 135 ms before contact, depending on the subject.

The studies by Whiting *et al.* (1970) and Lee *et al.* (1983) in fact seem to be marginally related to the point of how long it takes for visual feedback to influence movements. Strictly speaking, visual *feedback* must be the result of information generated by the moving segment. The above-mentioned experiments seem to relate more to *time to react* to a continuous change of the optical array during a movement of the target or during self-movement, than to visual feedback time (see Lee 1976). An elegant study in diving birds seems to stress this point. Lee and Reddisch (1981) have shown that retraction of the wings during the dive occurs some 60 ms before the bird comes into contact with the water surface. It is thus conceivable that a continuously changing stimulus will elicit more timely motor reactions than a suddenly changing stimulus, precisely because of the possibility of extrapolating the time to contact from optical parameters. In situations which involve unpredictable changes the time to react should lengthen. This is confirmed by the study of McLeod (1987) in professional cricketers. In this study the trajectory of the bat tip was analysed by high speed cine-film, in trials where the ball trajectory could be unpredictably altered as it bounced. Cricketers begin their shot long before the ball bounces. If the ball deviates from the expected trajectory after bouncing, a correction in the bat movement has to be made. McLeod found that no change in bat trajectory could be observed less than about 200 ms after the bounce. For the same reason, the time for processing visual feedback should also normally be longer in situations where neither the subject nor the target move with respect to each other, i.e. where the retinal information concerning target location is constant, which is the case in reaching for stationary objects. In other words there are restrictions to the Lee hypothesis for explaining short-term modifications or corrections of ongoing visually goal-directed movement.

3.2.4. Studies with saccadic eye movements

Eye movements directed at targets located within the peripheral visual field are usually composed of a main saccade, the amplitude of which undershoots target position by about 10 per cent, and a secondary, 'corrective' saccade, which brings the eyes on target. The latency of the secondary saccade is shorter than that of the main saccade. Typically an eye movement response to a 20° target step will be composed of an 18° saccade with a latency (T_1) of 250 ms, followed by a 2° saccade with a latency (T_2) of 130 ms (e.g. Bartz 1967). The difference in latency between the two saccades is not predicted by the classical Young and Stark (1963a and b) sampled-data model. This model assumed that the retinal error is sampled at a fixed rate of about five per second. Therefore, if the corrective saccade resulted from a new sampling of the retinal error at the end of the main saccade, and if sampling were synchronized on the end of the main saccade, the correction should be delayed by at least 200 ms. To account for the shorter delay observed experimentally, Becker and Fuchs (1969) proposed that the secondary saccade was 'pre-programmed' together with the main saccade, with the consequence that its computation and decision times were reduced. The main argument of these authors was that secondary saccades may be observed even when eye movements are made in the dark toward the position of previously learned targets. In the absence of actual targets, such secondary saccades cannot be considered as 'corrective' even though, as stressed by Becker and Fuchs, they are in the same direction as the main saccade and their latencies are similar to those of corrective saccades elicited by visible targets.

However, the hypothesis that the secondary saccade would be part of a pre-programmed package was questioned by Prablanc and Jeannerod (1975). These authors performed an experiment where eye movements were elicited by discrete target lights located at 5°, 10°, and 20° from central fixation. The target jumped from central fixation to one of the peripheral locations and remained on for 1 s (target step) or for a shorter duration (20–200 ms, target pulse). In the target-step situation, main saccades had a mean latency (T_1) of 217 ms. They were reliably followed by a corrective saccade (at least when target location was more than 5° from central fixation), the mean latency (T_2) of which was 190 ms. In the target-pulse situation, in which the target had been turned off before the main saccade was initiated, virtually no secondary saccades were observed (they were present in only 1 per cent of cases) and the undershoot of the main saccade with respect to target position remained uncorrected (Fig. 3.5). This point was later confirmed by Mather (1985).

This result is only apparently in conflict with that of Becker and Fuchs (1969). First, in their experiment the occurrence of secondary saccades in the absence of visual feedback was not, by far, the most frequent case. Second, the two experiments markedly differed in several ways. Although in the Prablanc and Jeannerod experiment the targets were turned off immediately prior to the movement, in the Becker and Fuchs experiment the target positions were first learned by the subjects by scanning the environment with their eyes. Then these targets were turned off and a series of movements were recorded in total darkness. It has been shown that in such a situation subjects tend to make large saccades, which overshoot the learned target positions (Jeannerod *et al.* 1965; Koerner 1975b). This fact was also clearly reported by Becker and Fuchs (1969). It might be that, in the prolonged absence of a visible target, the internal target image progressively lost its spatial relationship to the body reference and no longer guided saccades accurately. This internal target, however, was still used as a real target, and 'correction' saccades were made to approximate its position.

Another difference between the two situations was the amplitude of the target jumps (5–20° in Prablanc and Jeannerod, 10–90° in Becker and Fuchs). The importance of this point for generating different results was explored by Prablanc *et al.* (1978). They devised an experiment where targets jumped up to 50° from the midline and were turned off at the onset of the main saccade so that the eye movement was made in the dark. They found that, in cases where the main saccade undershot target position by 10 per cent or less, no secondary saccades occurred, thus confirming the

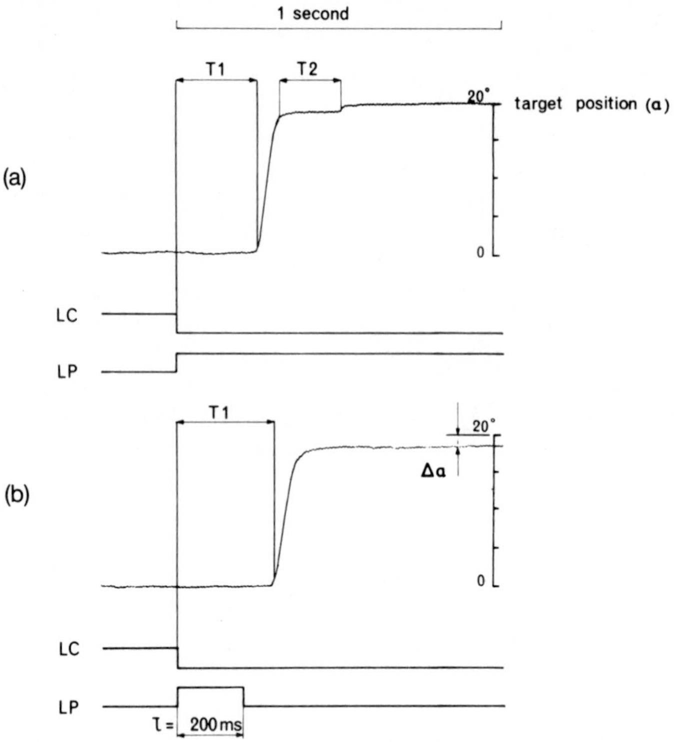

Fig. 3.5.   Saccadic responses in presence or absence of retinal feedback.

(a) Target step situation. A peripheral target (LP, 20° right) is turned on at the same time that the central target (LC) is turned off. A main saccade (photoelectric recording technique) is generated after a reaction time T1. The main saccade consistently undershoots target position and a secondary, corrective, saccade is generated after a time T2, shorter than T1.

(b) Target pulse situation. LP is turned on for a duration shorter than the reaction time T1. The main saccade is generated in the absence of visible target. In this condition, no secondary saccade occurs and the undershoot of the main saccade remains uncorrected. (From Prablanc and Jeannerod 1975.)

results of Prablanc and Jeannerod (1975). However, secondary saccades tended to be more frequent when the size of the undershoot of the main saccade increased, and were systematically present when the main saccade undershot target position by more than 30 per cent. Since eye movements in this experiment were performed in the absence of a visible target this finding tends to confirm the Becker and Fuch results.

In other experiments, Prablanc and Jeannerod (1975) investigated further the minimal requirements for triggering off corrective saccades by visual feedback. The results of these experiments are directly relevant to the problem of visual feedback time. A two-pulse situation was used. A main saccade was triggered by a short (20 ms) target pulse located 20° from fixation point. No secondary saccade was observed, unless a second 20 ms pulse was presented at the same location within about 50 ms of the end of the main saccade. In that case, a secondary saccade occurred with a corrective amplitude. Mean latency (T_3) of this corrective saccade with respect to the second pulse was $T_3 = 176$ ms (Fig. 3.6(a)). If the second pulse was presented with the same delay but at a location differing slightly from that of the first pulse (e.g. first target pulse at 20° from fixation point, second target pulse at 22°), the same result was obtained, namely, a corrective saccade was generated within a short period ($T_4 = 152$ ms; Fig. 3.6(b)). Finally, when the difference in location between the two target pulses exceeded a certain amplitude (e.g. more than 4°) the secondary saccade had a much longer latency ($T_5 = 275$ ms; Fig. 3.6(c)).

These results show that visual feedback is needed at the end of the main saccade for correcting eye movement amplitude and can be processed in a relatively short time, i.e. within the range 150–190 ms. The presence of corrections under certain conditions in the absence of visual feedback, as in the experiment of Becker and Fuchs (1969), indicates non-linearities in the correction mechanisms. One might suggest that when the error exceeds a certain value, it can be detected centrally by comparing the actual saccade size with the internal representation of the target position (Prablanc *et al*. 1978). Such a mechanism has been shown to be the basis for fast corrections (see Chapter 5).

It should be noted that results concerning visual feedback time obtained in eye movement studies may not generalize to other aspects of visuomotor behaviour. The short latencies of corrective saccades might represent a special case of visual reaction time. In eye movements the moving segment is not directly visible, and cannot be used as a source of visual feedback for approximating target position. The information for the correction is thus of the same nature as that for generating the movement itself, i.e. a retinal error signal. One can therefore speculate that attentional phenomena are gated by the main saccade, so that the response to retinal error is facilitated and the secondary saccade can be triggered within a short time.

3.3. Effects of altering visual feedback on reaching movements

The above sections have shown that under certain conditions visual feedback can account for regulation of goal-directed movements. First, the behaviour of the human operator in situations of altered visual feedback is compatible with that of well-defined control systems, particularly those involving intermittent sampling. Second, the temporal requirements for visual cues generated by a movement to influence the same movement are clearly met, except perhaps for very fast movements. The present section

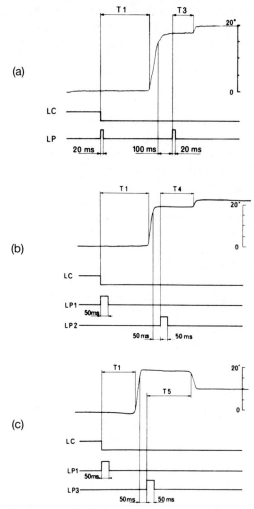

Fig. 3.6. Saccadic responses in the two-pulse situation.

(a) Two target pulses (20 ms) are presented at the same spatial location. The first pulse is sufficient to trigger the main saccade after the typical reaction time T1. The second pulse presented 100 ms after completion of the main saccade brings the eye at target location after a very short latency (T3).

(b) Two target pulses (50 ms) are presented at different locations. The first target pulse is presented at 20° on the right (LP1). It triggers a main saccade after reaction time T1. The second pulse is presented 50 ms after completion of the main saccade at 22° (LP2). It triggers an accurate secondary movement after a short latency T4.

(c) Same situation as in (b), except that the second pulse is presented at LP3 located at 10° from midline. The large retinal error between actual eye position after the first pulse and the second pulse (10°) generates another saccade after a long latency (T5). (From Prablanc and Jeannerod 1975.)

reports experiments where visuomotor cues were systematically suppressed or altered. The effects of these manipulations on movement accuracy and kinematics will both confirm the role of visual feedback in controlling movements and stress its limitations.

3.3.1. Methods for altering visual feedback

Suppression or alteration of visual feedback during execution of a movement has been, since Woodworth, a major experimental paradigm for the study of movement control. Different methods have been used. Total exclusion of vision can be achieved rather simply, either by asking the subject to close his eyes before he begins his movement, or by turning off the room lights. This method, used by the pioneer authors (and still in use nowadays), has the disadvantage of excluding not only vision of the moving limb, but also vision of the target and of the surrounding frame of reference. Therefore, conditions in which movements are executed with and without visual feedback are certainly not equivalent. A major technical improvement was introduced by Held and Gottlieb in 1958. Their technique, initially designed to study adaptation of eye–hand co-ordination to optical 'disarrangements' produced by prism spectacles, can also be used profitably for the study of normal eye–hand co-ordination. The main feature of the initial version of the Held and Gottlieb's apparatus (Fig. 3.7) was a mirror placed obliquely between subject's eyes and hand. The virtual image of the visual array thus appeared to project below the mirror. The mirror obscured the subject's hand while he pointed toward the targets (at the line crossings of the visual array). Marks made by the subject at or near the targets also remained invisible. Removal of the mirror restored a full view of the hand, hence allowing comparison between two relatively equivalent situations. The same principle can also be used for the presentation of solid three-dimensional target objects (see Jeannerod 1984—Fig. 6 in Chapter 2).

A more sophisticated apparatus based on the same principle has been built by Prablanc *et al.* (1979a) (Fig. 3.8). It will be described in some detail, since it has been used in several of the experiments to be reported in this and other chapters. Targets are presented through a matrix of eight red (600 nm wavelength) emitting diodes, which can be lit randomly. The subject sits, with his head fixed, in front of the surface R. He can see the virtual image E' of the target E, through a semi-reflecting mirror on the surface Q. When the space between surfaces Q and R is illuminated, the subject sees his hand through the mirror. When the light comes from above the mirror, vision of the hand is lost. Electronic shutters fed by logic pulses are used to control the lights.

Hand position is recorded on a surface covered with an isotropic resistive paper, fed by a current alternately switched along *x* and *y* axes (Bauer *et al.*

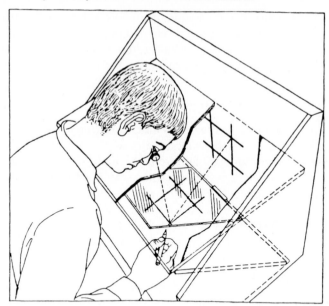

Fig. 3.7. The Held and Gottlieb apparatus for studying reaching movements without view of the moving limb. A mirror is intercalated between the targets (on the vertical panel) and the working plane (horizontal). If the mirror is at 45°, the targets appear to project on the working plane. The subject points at targets below the mirror without seeing his hand. (Reproduced from Held 1965, with permission.)

1969; Prablanc and Jeannerod 1973). A thimble is placed on the subject's forefinger. When the thimble is in contact with the recording surface, it acts as a cursor, the position of which is read alternately in x and y dimensions. The hand position signal is lost at the onset of the movement (when the hand leaves the surface) and a logic pulse is generated at this time; another logic pulse is generated at the end of the movement when the forefinger touches the surface again. The use of such an apparatus is not restricted to excluding visual feedback for the complete duration of the movement; opening or closing of the shutter also can be synchronized to the onset of the hand movement (or even the eye movement), or can be made at prescribed times after movement onset. The target also can be manipulated, e.g. turned off or displaced at the onset of hand or eye movements.

3.3.2. Effect of altering visual feedback on movement accuracy

Spatial and temporal characteristics of movements directed at visual targets and effected under visual control have been amply described. Some of these characteristics are predicted by Fitts' law that relates movement velo-

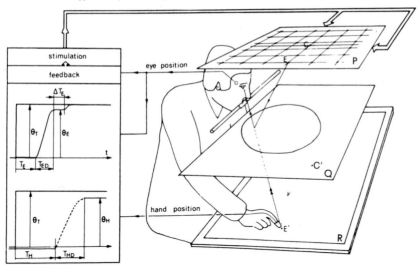

Fig. 3.8. Experimental apparatus for studying eye–hand co-ordination. The subject, with his head fixed, sees targets C and E presented on the matrix P through a semireflecting mirror Q. These targets project (C', E') on the working plane R. Turning on or off the light between Q and R allows or prevents vision of the limb, respectively. Latency, duration, and amplitude of saccades are recorded by an improved EOG technique (eye position). Latency, duration, and amplitude of hand movements are recorded from contact of the index finger with the surface R (hand position). Eye and hand signals can be fed into the computer and used for on-line modification of the target display or of the vision of the limb. (From Prablanc *et al.* 1979a.)

city and/or duration to distance of the target and to final accuracy. In the experiments by Fitts (1954) and his followers, movement accuracy was determined by the ratio of hits that fell within a given area surrounding the target, with respect to the total number of hits (hit and miss score). There are several problems with this method. First it only gives an estimate of movement accuracy and misses the information contained in the value of error amplitudes and in the scatter of pointing positions. In addition this method does not retain the sign of the errors and eventually misses a possible systematic bias in pointing positions.

(a) Accuracy measurements

Precise measurement of pointing positions with respect to the target position allows computation of three parameters of movement accuracy. The mean of pointing positions irrespective of target position represents the absolute error (AE). Together with the intrasubject standard deviation of the mean (sometimes itself called the variable error, VE), absolute error quantifies the scatter of pointing positions. The mean of pointing positions with respect to target position, obtained by retaining the directional sign of the error is the constant error (CE). Schutz and Roy (1973) have commented that AE is completely dependent on CE and VE and does not bring additional information. AE may be used as a global measure of pointing variability only if CE is equal to zero.

Some problems arise with interpretation of the significance of a constant error. CE is often observed in the amplitude domain, i.e. pointing movements in a given task or condition may systematically undershoot or overshoot target position, but the general direction of the movements is correct. On other (less frequent) occasions, provided the experimental design is appropriate, CE may reveal the existence of a systematic bias of the pointing positions toward a certain direction. This is the case if pointing movements are made away from the body, and pointing positions are consistently grouped too far to the right or to the left of target position. Following these remarks, it will be difficult, if pointing movements are effected in the frontal plane, to determine whether a constant error actually reflects a problem with encoding of the amplitude of the movements or of the position of the targets.

In normal conditions, the sign of CE in discrete pointing movements changes with movement amplitude. Subjects tend to overshoot proximal targets and to undershoot distal ones; in the midrange, constant error may be small or absent. This effect (the so-called 'range-effect': Brown *et al.* 1948; Slack 1953; Poulton 1980) is not related to absolute or specific target distances but, when observed, seems to be present within the range of distances used in each given experiment (for an example, see Fig. 3.9(b)). Another commonly observed effect if the relation of VE to distance of the target. As a rule, variable error increases as a function of movement amplitude (Brown *et al.* 1948; Poulton 1980). It should be noted that both the range effect and the increase in variable error with movement amplitude are more commonly observed in experiments using fast movements. In fact the procedure of speeding up the movements may somewhat obscure the issue of visual feedback control, because movements shorter than a certain duration will in any case be beyond the capacities of regulation of visual feedback, when present.

In situations where accuracy and not speed is emphasized, movement time is less prescribed by the index of difficulty of the task. Subjects tend to adjust the duration of their movements to some optimal value so that accuracy can be preserved for any distance or size of the target. This type of 'spontaneous' behaviour is constantly observed in saccadic eye-movements, perhaps because in this case the notion of target size and therefore of index of difficulty is largely irrelevant (targets for eye movements are defined in terms of spatial frequency components rather than in terms of absolute size). In other examples, lack of constraint on movement duration may even produce a time invariance effect where movements of increasing

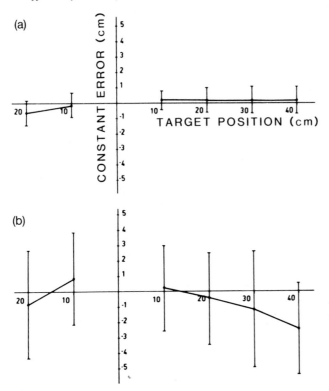

Fig. 3.9. Constant error (cm) and standard deviation (variable error) of hand pointing as a function of target position (cm) in normal subjects. Subjects used their right hand and pointed on either side of the midline (at 10, 20, 30, and 40 cm on the right, and at 10 and 20 cm on the left). (a) 'Visual feedback' condition. (b) 'No visual feedback' condition. Note large increase in variable error from (a) to (b). Also note constant error by undershooting for the more remote targets (range effect). (Data from Prablanc *et al.* 1979a.)

amplitude tend to keep a constant duration. Examples of this behaviour that clearly violate Fitts' law have been reported in Chapter 1.

(b) Effects of complete suppression of visual feedback

One advantage of studying movements executed without time constraints is that they are almost perfectly accurate when performed under normal visual control, which provides a baseline against which the effects of alteration of visual feedback on accuracy can be quantified. This methodology was used in the experiments of Prablanc *et al.* (1979a). In these experiments, subjects had to point accurately at targets located 10, 20, 30, and 40 cm from midline within a working space ipsilateral to the tested arm, and 10 and 20 cm from the midline on the other side. The apparatus shown in Fig. 3.8 was used. Movements were executed in a frontal plane 30 cm from the body, with the arm stretched, so that movements mostly involved the shoulder joint. Two conditions of visual feedback were used. In the 'visual feedback' condition, the hand was visible throughout the movement. In the 'no visual feedback' condition, vision of the hand was removed before, during, and after the movement. In both conditions the target remained visible for the duration of the trial. Targets were presented 10 times each in a random sequence. Measurements involved AE, VE, and CE, as well as latency and duration of the pointing movements.

As expected, virtually no CE was observed in the 'visual feedback' condition, and the variable error was very small (Fig. 3.9(a)). By contrast large pointing errors occurred when visual feedback was no longer available. The mean AE was correlated to movement amplitude with a positive slope. CE was also present: subjects overshot the position of targets close to the midline and markedly undershot the position of distant targets (the previously described range effect). Finally, VE increased with respect to the 'visual feedback' condition but was little influenced by distance of the targets (Fig. 3.9(b)). The fact that, with this type of movement, AE and CE increased with movement amplitude in the 'no visual feedback' condition while VE did not, is interesting to consider, because this dissociation might reflect distinct processes involved in movement accuracy. Increase in AE and CE is likely to reflect primarily the contribution of visual factors during the movement, which is conceivably more important in movements of a large amplitude. Global increase in VE also reflects partially the role of visual feedback. However, the fact that VE does not increase with movement amplitude in this condition as it does in conditions where movements are executed at maximum speed, indicates that this parameter is also related to motor factors, which are likely to become more critical when demands on the motor system are increased.

The types of errors observed in pointing at targets in the absence of visual feedback was also studied recently by Bock and Eckmiller (1986). In this experiment, subjects had to reach targets appearing in a random order by moving a lever around a vertical axis. The hand remained invisible during the movements. Constant error (by undershooting) was found to increase with target eccentricity, although variable error was little influenced, in accordance with Prablanc *et al.* (1979a). The working hypothesis of Bock and Eckmiller was that the programs of goal-directed movements encode movement amplitude rather than target positions. In order to demonstrate this point they devised another experiment where the targets were presented in a sequence such that the final position reached by one movement in the sequence became the initial position for the next. The results showed that CE and VE were larger for the last movements in the sequence than for movements closer to the departure position, due to the fact that errors made in each individual movement accumulated. This finding emphasizes the need for clearly dissociating mechanisms for encoding amplitude and direction of movements. As mentioned in the previous section, this would require specific experimental designs, a condition which was not achieved in the Bock and Eckmiller experiment. This problem will be fully discussed in Chapter 4.

In the experiment of Prablanc *et al.* (1979a) the temporal parameters of the movements were not affected by the presence or absence of visual feedback. The average duration of pointing movements directed at a 20° target, for example, was the same (*c.* 360 ms) with or without visual feedback. However, in either condition movement duration increased with target distance. An increase in duration of movements directed at targets of a constant size but located at increasing distances is predicted by Fitts' law. In the 'no visual feedback' condition, however, only one aspect of Fitts' law (the amplitude–duration relationship) seemed to be preserved. In this condition the size of errors was such that, in order to include 95 per cent of the hits as is normally required in this type of experiment, targets would have had to be exceedingly large, hence violating the relation between movement duration and index of difficulty.

The above considerations suggest that Fitts' law might not account for movements executed without visual feedback, at least beyond a certain level of task difficulty. This hypothesis seems to be confirmed by the results obtained by Wallace and Newell (1983). These authors have made a distinction between 'easy' movements (index of difficulty I_d below 4.5) and 'difficult' movements (I_d above 5). They found that easy movements had similarly small error rates whether they were executed with or without visual feedback. By contrast, difficult movements had much larger error rates when visual feedback was lacking. Even though in this case movement time remained closely correlated with I_d and Fitts' law was respected, one may wonder what significance should be given to such a relation in the condition of such a poor performance. This effect of task difficulty can be interpreted in terms of time for visual feedback to be effective: easy movements lasted less than 200 ms and therefore could well have been executed 'open-loop' in either 'visual feedback' or 'non visual feedback' conditions. A better interpretation (Wallace and Newell 1983) would be to assume that the relation between movement duration and index of difficulty is independent of visual feedback and pertains to the program of the movements (see above).

Removing vision of one's hand during movement toward a target is by no means the only way of altering visual feedback. There are other experimental conditions, which produce decorrelation of efferent (motor) signals from the normal reafferent (feedback) signals. Smith *et al.* (1960) have shown that such a decorrelation, by way of a delay introduced in a closed TV circuit, produces a considerable alteration in simple visuomotor tasks like drawing. Effects of delayed visual feedback on goal-directed movements have been reinvestigated more recently by Smith and Bowen (1980). In their experiment, subjects had to point with a pen at a target within fixed durations between 150 ms and 650 ms until they reached the criterion of four successive movements within ± 40 ms of the required time. Movements could be performed under three different conditions, e.g. normal visual feedback, delayed visual feedback (delay was 66 ms), or prismatic displacement of 10°. In the two conditions where visual feedback was either delayed or distorted, the measures were made on the first three movements performed at duration criterion. In the normal visual feedback condition the hand was found to undershoot target position for the movements executed in 250 ms or less, although it was accurate for longer durations. In the prism situation a large constant error (by overshooting, as expected from the direction of prism displacement) was also observed in the fastest movements. Finally, in the delay condition, movements were very inaccurate and overshot target position. This effect was particularly clear for longer movement durations—it may be explained by the late arrival of the corrective feedback signals. This result indicates that in order to be effective, visual feedback signals must be tightly correlated in time with other signals arising from the movement.

3.3.3. Effects of altering visual feedback on movement kinematics

It has been known since Woodworth that movements directed at visual targets include one main component covering most of the distance to the target, and a final adjustment. Woodworth claimed that the final adjustment was due to visual feedback, because it disappeared when movements were executed with the eyes closed or at a fast speed, two conditions where he logically thought visual feedback was excluded.

(a) The kinematic pattern of accurate movements

Woodworth's observations have been confirmed by the many authors who have been able to directly record movement trajectories. It is true that the basic pattern of aiming movements changes with the nature of the task. Movements performed at a maximal speed have an acceleration phase (up to the point of maximum velocity) and a deceleration phase of about equal duration (Peters and Wenborne 1936). 'Free' movements executed without target or directed at very large targets, or in general movements without a high accuracy requirement, also tend to have a symmetrical velocity profile (Crossman and Goodeve 1963; see also Abend *et al.* 1982; Hollerbach and Flash 1982; Lacquaniti *et al.* 1982). By contrast, movements approaching a small target and requiring accuracy have velocity profiles which are clearly asymmetrical, with a deceleration phase longer than the acceleration phase. The difference between movements with and without a precise goal is thus interesting since it demonstrates that the longer deceleration observed when a high accuracy is required must be due to an active process for bringing the movement at the target, rather than to passive counteracting forces.

A longer deceleration implies that a certain level of velocity is maintained in the late part of the trajectory. In other words a movement directed at a small target would unavoidably undershoot target position if it had a symmetrical velocity profile. Now the problem is to determine the nature of the additional displacement which complements the initial large movement, i.e. is it a smooth change in velocity or is it composed of one or several secondary re-accelerations. This is a critical issue because of its implication on mechanisms postulated for movement control. In fact, the significance of the asymmetry of the velocity curve may be difficult to interpret when the moving segment has a non-neglegible inertial load, like the arm. In that case, the apparent smoothness of the deceleration curve could be due to the filtering out of smaller re-accelerations during the deceleration phase. The occurrence of small sub-movements at the end of an aiming movement has been postulated within the frame of intermittent feedback models ('micro-ballistic') corrections (Craik 1947; Crossman and Goodeve, 1983; Keele 1968; see also Section 3.1.2). This type of motor organization implies the existence of a main motor impulse accounting for the initial acceleration phase, followed by smaller correction impulses producing brief re-accelerations. An alternative explanation for assymetrical velocity curves is that the deceleration phase would be composed of a smooth low-velocity movement driving the moving segment to a progressive stop at the desired position. In that case, the initial impulse producing the ballistic movement would have to be followed by a 'ramp' movement, i.e. a type of movement permanently controlled by some feedback signal.

There are definitely more arguments in favour of the double (or multiple) impulse organization. These arguments are drawn from experimental studies of eye and arm movements. Saccadic eye movements are interesting in this respect because, due to the small inertial load of the eye, no damping occurs, so the secondary movements remain visible. Secondary movements in fact seem to represent a basic organization within the oculomotor system. An experiment by Henson (1978) demonstrates this point very elegantly. This author fitted normal subjects with a diminishing optical device. The effect was such that a target appearing, say, 30° to one side of the fixation point required an eye movement of less than 30° to obtain fixation. It was predicted that after prolonged exposure to such a condition subjects should learn to make a single saccade in fixating targets since secondary saccades would no longer be required to correct undershooting of the target by the main saccade. In fact, it was found that after a few minutes, the main saccades again undershot target position as they had before visual feedback was altered by optical diminutions and the second-

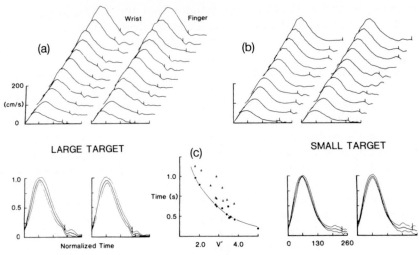

Fig. 3.10. Temporal characteristics of arm movements to a target. (a and b), Velocity profiles of wrist and finger during pointing at a large target (a) and a small target (b). Upper row: individual trials ordered according to wrist velocity. Duration of movements has been normalized. Lower row: averaged velocity profiles of the same movements + or − one standard deviation. The small vertical bars indicate time of contact with target. Note longer deceleration time and occurrence of secondary movements in the small target condition. (c) variation of movement time with speed. Black dots: movements to the large target. Triangles: to the small target. (From Soechting 1984, with permission.)

ary saccades reappeared. Henson's conclusion was that undershooting is a deliberate mechanism of the saccadic system. This statement seems to be verified for almost any saccade of an amplitude beyond 5° executed in conditions of visual feedback.

It is tempting to generalize this hypothesis to other aspects of visuomotor behaviour. Concerning reaching arm movements, recent work involving precise analysis of arm trajectories toward a target have revealed the existence of discrete secondary movements. In a study by Soetching (1984) the trajectory of the finger was recorded during pointing movements directed at targets, either large (5 cm) or small, located at 35 cm from resting position. The velocity profiles of the movements were reconstructed by sampling every 20 ms the position of an ultrasound-emitting source placed on the fingertip. It was observed that, whereas the acceleration phases were highly stereotyped across movements, including movements directed at targets of different sizes, the deceleration phases were much more variable (Fig. 3.10(a)). When movements were aimed at small targets, duration of the deceleration phases was prolonged by comparison with movements aimed at larger targets. The target was approached at a low velocity and in many instances the velocity profile showed a secondary peak before contact was made with the target (Fig. 3.10(b)). These features were observed for movements of different velocities.

Secondary movements were also observed by Jeannerod (1984) in a reaching task where subjects had to grasp small objects placed in front of them. It was found by reconstructing the velocity profiles of these movements that the deceleration phases were consistently marked by a discontinuity, where the arm velocity tended to become constant for a short duration, or even to re-increase before deceleration was terminated and the movement stopped. The fact that secondary movements occur more systematically in grasping than in pointing may be explained in a simple way. Accurate reach during grasping can be obtained only through an exact planning of the amplitude of the movement. By contrast, the control of amplitude is not critical in pointing movements, where accuracy is compatible with a relatively fast approach and with hitting the target. Secondary movements, which are related to the control of accuracy in the amplitude domain, are thus more likely to be observed in grasping than in pointing (see also Marteniuk *et al.* 1987).

Another situation where secondary movements are consistently observed is that of micromovements. Unpublished observations made by Jeannerod with D.N. Lee and C. Trevarthen from Edinburgh showed that very small reaching movements executed during work under a microscope were composed of several sub-movements. In this experiment subjects reached with a small stylus for targets placed a few mm apart, and viewed through a binocular microscope (magnification × 40). The tip of the stylus was filmed at 50 frames/s. Typically, these micromovements lasted between 200 and 400 ms. Their velocity profile showed a sharp velocity peak, consistently followed by one or several secondary movements (Fig. 3.11). The secondary movements occurred within a relatively short time after the peak of velocity of the initial movement, which suggests that they were independent from visual feedback.

(b) Are secondary movements dependent upon visual feedback?

Observations on different types of reaching movements, together with those on eye movements (see Section 3.2.4) indicate that secondary movements might be an ubiquitous phenomenon, related to systematic undershooting in aiming tasks when terminal precision is required. Thus it is interesting to look for the effect of alteration of visual feedback on the occurrence of these secondary movements. This is a controversial issue, first because of discrepancies between results from different authors, and second because of the implication of these results for the underlying hypothesis about movement control. Woodworth's claim on this point was that when it was impossible to use visual feedback during movements, the result was lack of final adjustment of the trajectory. Final adjustment, according to Woodworth, could be effected through a slowing down of the

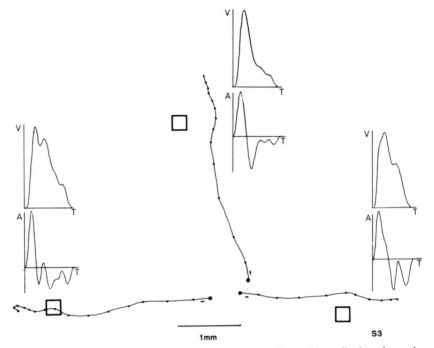

Fig. 3.11.   Kinematic pattern of micromovements. The subject displaced a stylus from a starting position (black circles) to targets (squares). The targets and the stylus were seen through a dissecting microscope (magnification × 40). The movements of the stylus were filmed at 50 frames/s by using the same magnification. Trajectories were reconstructed from the films (arrow indicates direction of movement). Corresponding tangential velocity (V) and acceleration (A) profiles were constructed.

Movements lasted about 200 ms (each dot on the movement trajectory represents 20 ms). They frequently overshoot target position, which may be explained by visuomotor conflict between the actual size of the display and its appearance through the microscope. Note several secondary movements in the final part of the trajectories. (Unpublished data by Jeannerod, Lee, and Trevarthen.)

movement near its end or by way of small extra movements. Disappearance of secondary movements in conditions where visual feedback is excluded would thus represent a simple explanation for the inaccuracy and constant error (usually by undershooting) of reaching movements executed in this condition. The problem, however, cannot be solved in such a simple way.

The problem of the nature and the role of secondary movements in accurate aiming tasks was addressed by Meyer *et al.* (unpublished) in an experiment involving reaching for targets of variable widths and of various distances displayed on a CRT screen, by rapidly rotating a handle. The angular position of the handle was monitored by a cursor appearing on the same screen as the targets. Visual feedback from the movement was manipulated by turning the cursor on or off (visible and invisible cursor conditions, respectively). In either condition, total movement duration was highly correlated to both target width and distance, as predicted by Fitts' law. Error rate (percentage of hits falling outside target boundaries) also increased linearly with index of difficulty. Shifting from the visible to the invisible cursor condition increased the overall error rate without affecting its relation to index of difficulty.

Secondary movements were observed more frequently in movements involving a high index of difficulty (they were present in up to 96 per cent of trials) than in easier movements, but their frequency was never below 40 per cent. Interestingly, secondary movements occurred at the same rate whether the cursor was visible or not. However, in spite of not being influenced by visual feedback, these corrective movements made an important contribution to the relation of total movement duration to index of difficulty. In fact, secondary movements accounted entirely for this relation, since duration of the first movement alone was not correlated to index of difficulty.

These results have important implications. On one hand, they tend to confirm the iterative correction model (Crossman and Goodeve 1963; Keele, 1968) implying that Fitts' law relies mostly on 'corrections'. On the other hand, however, in showing that secondary movements owe nothing to vision, they tend to discredit the visual feedback hypothesis of Fitts' law. Meyer *et al.*'s interpretation was that optimization of the submovements that compose a reaching movement pertains to the program that controls the movement, not to feedback from its execution, according to their previous impulse variability model (Meyer *et al.* 1982, see above).

In the previously mentioned study by Jeannerod (1984), reaching movements of the hand were also studied in the absence of visual feedback from the moving limb, by using the apparatus shown in Fig. 2–8. Analysis of the velocity profiles of these movements showed that the discontinuities in velocity observable during the deceleration phases were still present in the absence of visual feedback. This fact, which holds true whether the target-object remains visible throughout the movement, or the lights are turned off at movement onset, seriously departs from the Woodworth's model. It must be remarked however that Woodworth had inferred the disappearance of final adjustments only in the case of movements executed rapidly, which may represent a special case.

Persistence of secondary hand movements in the absence of visual feedback seems to be in direct contradiction to the type of behaviour observed in eye movements by Prablanc and Jeannerod (1975) and Mather (1986) (see above, Section 3.2.4). In fact there are many differences between hand and eye movements. Although eye movements executed without visual feedback are probably devoid of other sources of reafferent control this is not the case for hand movements, where kinaesthetic feedback is of a major importance. Therefore it could be suggested that secondary movements following the main acceleration in hand reaching are at least in part dependent on proprioception. This point will be discussed in another chapter. Another major difference between the two categories of movement is the number of degrees of freedom involved. It is hard to believe that the many degrees of freedom and the complexity of muscular commands involved in a hand-reaching movement could all be handled by visual feedback alone. Visual feedback relates to the endpoint of the limb, but does not reflect the intrinsic aspects of the movement which have to be controlled for achieving accuracy. By contrast a simple terminal feedback may be sufficient for controlling accuracy in ocular saccades. This difference might well explain why secondary movements are more dependent on visual feedback in eye movements than in arm movements (see Jeannerod 1986b).

It remains to be determined at which level of movement control vision acts to improve accuracy. Surprisingly, there are very few data concerning this point and very few investigators seem to have attempted a comparative kinematic analysis of movements executed in conditions where visual feedback is present or excluded. Carlton (1981) made high speed cine-film analysis of movements executed under different degrees of visual feedback exclusion. The technique used by Carlton implied that the initial part of the hand movement toward the target could be masked, but that terminal feedback was always available. The main finding related to the present discussion was that movement time increased as a function of the amount of movement masked. The lengthening of movement time was due to a longer time spent near the target, although the initial part of the trajectory had a similar time course whatever the amount masked. This result tends to indicate that lack of visual feedback would slow down the deceleration phase, by maintaining low velocity for a longer time before the stop. This effect however might not be due to lack of visual feedback *per se*, but rather to a systematic strategy of the subject of waiting until terminal feedback becomes available. Another of Carlton's findings was that correction movements were generated a short time after visual feedback became available. Again, this does not mean that secondary movements as those described earlier in this chapter would be triggered by vision. We have seen this not to be the case for hand reaching. Secondary movements observed in the Carlton's situation might also be part of the same strategy of 'waiting' for vision of the terminal error.

3.4. The nature of visual signals for feedback control

The term 'visual feedback' is a very broad one. In fact, vision of one's own limb moving toward a target generates several different cues, which might contribute in different ways to the global improvement of movement accuracy by vision. Experimenters have attempted to alter these cues separately. This is a difficult task, however, and most of their experiments have remained inconclusive because of the impossibility of effectively eliminating one factor without affecting the others.

3.4.1. Central and peripheral visual channels

The question of whether signals arising from the moving limb and used to correct movement trajectory, are dynamic (velocity) or static (position) signals can be raised first. This question seems justified by the fact that each of these two categories of signals may be encoded by different visual channels. As far as only visual cortex is concerned here, it has been shown to contain directional neurons that are tuned to different velocities of moving stimuli. In addition these neurons are selectively distributed within areas of visual cortex controlling different zones of the visual field. Experimental data in animals provide interesting information on this point. In the cat, cells that respond best to low velocities of stimulus motion ($0.5–5°$ s^{-1}, velocity low-pass cells) strongly dominate in that part of area 17 that subserves central vision ($0–10°$). In areas subserving peripheral vision, particularly in area 18, most of cells respond best to high velocities ($10–500°$ s^{-1}, velocity high-pass cells) (Orban *et al.* 1981) (Fig. 3.12(a)). Selectivity of repartition of these cell types according to retinal eccentricity seems quite remarkable. In addition, some cells in the visual cortex are tuned for a narrow range of velocities (velocity-tuned cells) and therefore detect best certain types of movement. These cells are mostly found in areas subserving central vision (Orban *et al.* 1981) (Fig. 3.12(a)). A similar organization has also been observed in the monkey. In the part of primary visual cortex subserving the central visual field neurons respond best to low velocities, and in that part subserving the peripheral field, they respond best to high velocities (Orban *et al.* 1986) (Fig. 3.12(b)).

In humans, study of motion thresholds has also provided evidence for discontinuities across different retinal areas. In the central part of the visual field, very small displacements can be detected, as could be predicted from the high spatial resolution of central retina. In addition, introduction of a stationary mark in the motion path of a moving object has been shown to further decrease this detection threshold in the central retina, though not in the peripheral retina (Tyler and Torres 1972). These mechanisms would account quite well for detection by the central retina of errors between the moving hand and a stationary target. Accordingly, suppression of the sight of the target during a pointing movement has been shown to severely degrade pointing accuracy, even in the case where the hand remains visible (see Chapter 5).

Velocity detection in humans has been shown by several authors to be

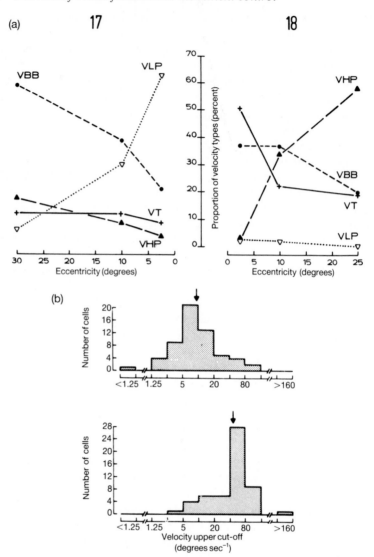

Fig. 3.12.   Response of cortical cells to velocity.

(a) Proportion of the four different types of cell responses to velocity as a function of eccentricity in areas 17 (left) and 18 (right) in the cat. Zero degree indicates
centre of visual field. VT: velocity tuned cells. VBB: velocity broad band cells.
VLP: velocity low-pass cells. VHP: velocity high-pass cells. (From Orban *et al.*
1981, with permission.)

(b) Distribution of velocity upper cut-off of cells in area V1 in the monkey.
Upper diagram, V1 central (0° to 2°). Lower diagram, V1 peripheral (15° to 25°).
Note low-velocity cut-off in the central area, and high-velocity cut-off in the peri-
pheral part. (From Orban *et al.* 1986, with permission.)

equally effective in all parts of the visual field (Barbur and Ruddock 1980; Bonnet 1982). More recently, however, Orban *et al.* (1985) have shown a gradual shift to detection of higher velocities with increasing retinal eccentricity of the moving stimulus. This result therefore indicates that visual motion signals are not processed in the same way across the visual field and that, as in the cat or the monkey, the peripheral visual field is more sensitive to higher velocities than the central visual field.

3.4.2. Relevance to visuomotor control

Although this view of visual system functioning may be a simplified one, it bears some value for understanding visual control of movement. When a subject reaches for a target by hand, he first orients his eyes at the target location in order to achieve foveal fixation. It is known that the sequence of movements is such that foveal fixation may occur before the hand has even started to move (Prablanc *et al.* 1979a; Biguer *et al.* 1984; see Chapter 2). Therefore during the first part of its trajectory (i.e. when its velocity is higher) the hand will be seen through the peripheral visual field, and the final phase of the movement where the hand moves at a much lower velocity, will be seen through the foveal field.

It is thus easy to find a correspondence between the duality of visual detection mechanisms for motion signals and the duality of visuomotor mechanisms of reaching movements. The low-velocity final phase of the movements would be controlled by visual feedback signals originating from the foveal field, through the velocity-low pass or velocity-tuned cell-types. Indeed, it has been shown repeatedly that visual feedback from this part of the movement is critical for terminal accuracy (e.g. Carlton 1981). Poor accuracy of movements executed at a fast speed, i.e. in cases where hand velocity during the final phase exceeds the capacities of detection of the cells in the foveal field, would be explained by functional exclusion of this mechanism.

The high-velocity initial phase of the movements would also be submitted to visual feedback control. However, it is difficult to determine which parameter it would be most appropriate to detect at this stage of the movement. *Peak velocity* is a strong index to movement amplitude, though it is doubtful that it could be detected because of the small numbers of velocity-tuned cells in the cortical areas subserving the peripheral visual field. Detection of movement *direction* would also be very advantageous, principally because this type of feedback is available almost immediately after the movement has started. Also, early detection of directional errors would allow the corresponding corrections to be minimal and in any case much smaller than if the same directional errors had to be corrected at the end of the movement. The fact that feedback about movement direction can be sampled and used in movement control is suggested by the experimental results of Zelaznik *et al.* (1983) and Elliott and Allard (1985), who have shown that in very fast pointing movements the presence of visual feedback reduces directional errors, but not distance errors.

There are several possible ways to test the hypothesis of duality of visuomotor channels for feedback control. One way is to ask subjects not to move their eyes when they are pointing at a target appearing in the peripheral visual field. In such a case, fixation of the target by the central retina would not occur and movement of the hand would be effected under the sole control of the peripheral retina. This experiment has been done by Prablanc *et al.* (1979a) with the result that accuracy of pointing appears to be severely degraded. Such a finding, however, cannot be interpreted only along this line of reasoning. Preventing ocular fixation of the target not only suppresses foveal control of the terminal phase of movement: it also alters cues, related to the position of the gaze, which may be important for the encoding of target position (see Chapter 4).

Another test of the same hypothesis is to suppress velocity signals from the hand movement, e.g. by stroboscopic illumination. Preliminary results mentioned by Paillard (1982) are relevant to this point, although they were obtained in a prism adaptation experiment and not in an experiment where accuracy of pointing was directly measured as a function of available cues. In this experiment the adaptation phase was made under stroboscopic light. Two procedures were used for adaptation. When the procedure described by Held and Freedman (1963) was used (i.e. subjects moved their hand back and forth on a neutral background while they watched it through the laterally displacing prism), no adaptation occurred. When the procedure used for adaptation involved attempts to reach for targets with the adapting hand, a larger adaptation was observed. Paillard's interpretation of these results was that adaptation could not occur in the first situation because it relied only on the velocity cues, which were abolished during stroboscopic illumination. By contrast adaptation occurred in the second situation because it relied on position cues, which were preserved during stroboscopic illumination (see also Paillard *et al.* 1981).

Whatever the respective roles attributed to central or peripheral vision for the control of direction, velocity, or amplitude of movements, these two visual modes are clearly subserved by different anatomical and physiological substrates. It is therefore important to study visuomotor performance in conditions where specific parts of the visual system are lesioned or excluded. This approach is the basis for the concept of duality (or multiplicity) of visual systems (Trevarthen 1968; Schneider 1967). The next section will show that visuomotor function can be clearly dissociated from other visual mechanisms, and can be mediated in the absence of the visual cortex.

3.5. Effects of visual cortex lesions on reaching movements

Pathological destruction of the visual cortex in humans is classically thought to produce total blindness, except for pupillar response to light and very crude visual perception limited to sudden changes in illumination—a condition known as cortical blindness. This conventional opinion, however, has been called into question, largely on the basis of experimental findings in monkeys. Although destriated monkeys appear to be profoundly impaired in their ordinary visual behaviour, they are still able to generate motor responses directed at objects appearing in, or moving across, their visual field (Humphrey and Weiskrantz 1967; Feinberg *et al.* 1978). These residual visual abilities have been attributed to activity of subcortical structures surviving ablation of the visual cortex, among which superior colliculi seem to play a critical role (see Mohler and Wurtz 1977).

In humans, evidence for monkey-like residual visual abilities following lesions of the striate cortex was first suggested by the experiments of Poeppel *et al.* in 1973, followed by those of Weiskrantz *et al.* in 1974 and Perenin and Jeannerod in 1975. The common feature in these experiments was that they used a new methodological approach derived from the monkey studies. In humans, visual functions are commonly assessed by asking the subject whether or not he can 'see' the stimulus. Patients suffering from a unilateral lesion of the visual cortex do not report seeing targets within the contralateral half of their visual field, i.e. they lack the common subjective experience of 'seeing'. By contrast, in monkeys whose subjective experience is not accessible to the experimenter through language, visual abilities are tested in a different way. Animals are trained to reach for targets or to indicate manually which of two patterns they have selected. Poeppel *et al.* (1973) were the first to suspect that such behavioural responses could also be useful for testing visual capacities in humans. Indeed, they asked their cortically lesioned subjects not to try to see stimuli that were presented within their scotomata, but rather to try to locate them by turning their eyes toward them. Poeppel *et al.* recorded eye movements, the direction and amplitude of which were weakly but definitely correlated with position of the targets. Because the subjects remained unaware of the stimuli, they subjectively experienced 'guessing' rather than 'seeing'. In other words, demonstration of 'blindsight' (a term introduced by Weiskrantz and his colleagues, see Sanders *et al.* 1974) could be obtained when subjects were required to abandon their usual, perceptual mode of visual detection, and to use a visuomotor mode based on a forced motor response.

It is interesting to examine in more detail the motor responses obtained under such conditions and to compare them with those obtained from normal subjects. Data for this comparison will be drawn from the study of residual vision in a group of six hemispherectomized subjects operated at ages ranging between 6 and 9.5 years in five cases and at 17 years in one case (Perenin and Jeannerod 1978).

3.5.1. Pointing accuracy within the hemianopic field

In this experiment, subjects were placed at the centre of a 57 cm radius semi-cylindrical vertical screen, diffusely illuminated in the mesopic range. Targets were projected on the screen at the subject's eye level at 15, 30, 45, 60, 75° on each side of the midline. Bright stimuli (a 6 × 8° horizontal rectangle, or a 2–3.5° diameter spot) presented for 100 ms or 500 ms were used as targets. Subjects, with head movements restrained, had to binocularly fixate a mark placed at the centre of the screen. Central fixation was monitored by the experimenter by way of a television system. An acoustic warning was given prior to each stimulus presentation. The subject's task consisted in pointing rapidly with the arm at the target location, while keeping the gaze fixed at the centre of the screen. This task was easily achieved for targets appearing within the normal half of the visual field, where the subjects could see the targets. For targets appearing on the hemianopic side, however, subjects were required to point immediately after the acoustic warning, at the point where they 'guessed' the target had appeared. Because only the arm ipsilateral to the hemispherectomy could be used for pointing, subjects had to hold a 25 cm stick in order to be able to reach for the most peripheral targets within the hemianopic field. Errors were measured to the nearest 1° by reading the pointing position on a scale placed directly on the screen.

Results from this experiment, reported in detail by Perenin and Jeannerod (1978), clearly showed that subjects did not point at random: pointing positions within the hemianopic field were definitely correlated with target positions. In the sessions where targets were presented for a short duration (100 ms), correlation coefficients significant at the 0.001 level were obtained in three subjects. In one subject, the correlation coefficient was significant at the 0.05 level; in another one, it was non-significant; finally, one subject was not tested in this condition. For target presentations lasting 500 ms, correlation coefficients were significant in all cases.

Larger constant errors were observed in all sessions. Subjects usually overshot target position for targets located within 30° of the midline, although they undershot targets located beyond that eccentricity. This is demonstrated in Fig. 3.13, which displays combined results from the six patients. For the most peripheral targets (e.g. beyond 60°) undershoot might have been partly accounted for by mechanical factors, since subjects had to use their arm contralateral to their hemianopic field. It remains, however, that even at 45° an undershoot as large as 10°–15° could be observed. Increase in the duration of presentation of the targets (from 100

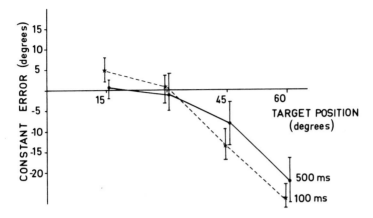

Fig. 3.13. Constant and variable errors in pointing movements executed within the hemianopic field in six hemidecorticated patients, as a function of target position. Dashed-line: 100 ms exposure of the target. Continuous line: 500 ms exposure. Data presented in Fig. 3(a) and (b) of Perenin and Jeannerod (1978) have been replotted.

to 500 ms) tended to reduce the amount of this constant error (Fig. 3.13). Large undershooting for the most eccentric targets has also been observed in experiments where targets were localized by saccadic eye movements, instead of by arm pointing (Weiskrantz *et al.* 1974; Zihl 1980).

3.5.2. Comparison of pointing performance in normal and hemianopic subjects

Demonstration of relatively accurate pointing in hemianopic fields has a number of important implications for the mechanisms that control both the initiation and the accuracy of goal-directed movements. The large errors made by cortically lesioned subjects in pointing toward their hemianopic field must be compared with those made by normal subjects. This comparison, however, can only be done with normal subjects pointing in a situation where control cues for the movement are reduced by the same amount. Examination of blindsight implies no conscious visual feedback from the moving hand, no reference signal from eye or head rotations toward the target (the subject has to fixate the centre of the screen), and no information about target location during the movement (target exposure was limited to 100 or 500 ms in our experiments). Such a situation, which combines the difficulties of the 'no visual feedback' condition with those related to the lack of target information during the movement and the lack of eye position information, has been shown to result in large errors in normal subjects (see Chapter 4). If the comparison is made for targets of the same

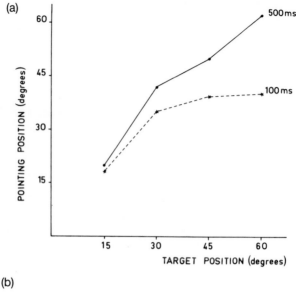

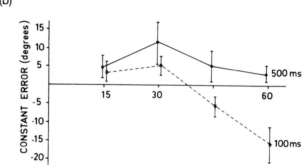

Fig. 3.14. Pointing performance in a hemianopic subject with a congenital poren-cephalic cyst of the posterior part of the left hemisphere. (a) plot of pointing pos-itions as a function of target positions within the hemianopic field. (b) constant and variable errors during the same pointing sessions. Dotted lines: 100 ms exposure of the targets. Continuous lines: 500 ms exposure. (From Hécaen *et al.* 1984.)

eccentricity in both normal and hemianopic subjects, one finds that in the two groups the direction of the errors is similar (i.e. overshoot for targets close from the midline and undershoot for more remote targets), and that the size of errors is in the same order of magnitude (compare Figs 3.13 and 3.14 with Fig. 3.10). The general pattern of responses to stimuli presented in the hemianopic field of subjects with cortical lesions can therefore be regarded as an exaggeration of the pattern of response of normal subjects.

This conclusion has an important consequence with regard to what has been said in the previous two chapters. It indicates that movements can be initiated without a direct cognitive representation of the goal. This is an example of the functioning of discrete neural channels subserving discrete behavioural productions. The lesion artificially splits behaviour into modular compartments (in this case a cognitive level and a visuomotor level) which normally cannot be dissociated (Jeannerod 1981a). The fact that cognitive experience is disconnected from information transfer and motor production mechanisms does not mean, however, that visuomotor channels make a direct transformation of visual input into motor output. The work of Zihl and Werth (1984a, b) has clearly shown that hemianopic subjects, even though they remain unaware of their pointing performance, can be trained to better pointing. Their two patients were required to direct their eyes where they guessed the targets had appeared. They initially tended to make saccades of a rather constant amplitude, without reference to target location. They were then informed that targets would appear at a different location each time, and that they should therefore shift their eyes by a corresponding amount. They were never informed, however, about the location of the target, nor about their localization performance. This 'shaping' procedure, similar to that used in monkey training experiments, produced in the patients a clear and rapid improvement in performance (Zihl and Werth 1984b). Thus, the internal 'pre-motor' processes related to generation of visually directed movements can be accessed on a non-cognitive mode. The nature of the task (e.g. cognitive or not) selects a given mode of functioning and, in consequence, a given visuomotor channel.

3.5.3. Control experiments

Blindsight is a controversial issue. It has been suggested by its detractors that results such as those of the above-mentioned pointing experiments could be artefactual because subjects could use light scattering from the targets into unimpaired parts of their field as a localizing cue (Campion *et al.* 1983). There is no doubt that scattered light may be a problem in such experiments, particularly when bright targets and a high target to background contrast are used. However, the amount of blindsight has been found to vary across groups of patients examined in the same experimental situation (Hécaen *et al.* 1984; Perenin and Jeannerod 1975). This fact clearly indicates that because differences between groups cannot be related to experimental conditions, they have to pertain to specific characteristics of the groups, such as age, or type or extent of lesion.

This problem of poorer blindsight performance in certain patients, particularly in cases with lesions occurring late in life, is quite apparent in the study of Campion *et al.* (1983). These authors have reported on pointing performance of three hemianopic patients with lesions of the occipital cortex acquired in adulthood. Stimuli used as targets were as bright as those

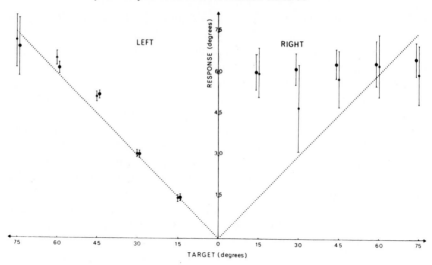

Fig. 3.15. Pointing performance in two subjects with hemianopia of chiasmatic origin. Results from the two subjects have been averaged. The target was a 8° square pattern, exposed for 100 ms (small dots) or 500 ms (large dots). Left: normal hemifield. Right: hemianopic field. Note accurate pointings in the normal field. Within the hemianopic field, subjects pointed at arm length, irrespective of target position. (From Perenin and Jeannerod 1978.)

used in our experiments (Perenin and Jeannerod 1975), and were presented for a longer duration (1 s). Accuracy of pointing at targets presented within the scotoma was very poor in all three subjects, correlation between target position and pointing position being clearly significant in one case only, non-significant in one case and marginally significant in one case.

The same criticism of scattered light cannot apply to another control experiment performed by Perenin and Jeannerod (1978, 1983) in two adult patients with a bitemporal hemianopia resulting from chiasmatic lesions. In such cases no blindsight is expected to occur, since all retinal projections subserving the temporal fields are blocked by the lesion. Indeed in these two patients, positions of pointings directed at targets appearing within the temporal scotomata were found to be unrelated to target positions (correlation coefficients ranging between 0.09 and 0.37, non-significant) (Fig. 3.15). This result indicates that scattered light could not be used by these patients for localizing the targets. Scattered light is an unspecific factor common to all hemianopic subjects examined for blindsight. Therefore, there is no reason why it should not be used by all of them for directing the movements at target location. By contrast, blindsight is a delicate effect revealing the residual activity of an impaired system, and there is no reason why it should be present in all hemianopic patients with cortical lesions.

3.6. Conclusion

The experiments reported in this chapter clearly demonstrate that visual feedback can influence the execution of movements. Its role might be mainly to control final accuracy by using signals related to the position and the velocity of the end-point of the limb. The hypothesis that position signals might be preferentially processed by the central visual field, while velocity signals would be processed by the peripheral field, is not entirely supported by the experimental data.

However, the view of visuomotor control as a closed-loop mechanism is at least partly incorrect. Relatively good performance can be preserved in movements executed in the absence of visual feedback (provided they are not executed at maximal speed). Indeed, vision may control movements by other ways than by visual feedback. It is conceivable that the visual 'map' (or representation) of space is also used for correcting errors in executing movements. This action would be exerted in conjunction with non-visual feedback signals, generated by kinaesthetic receptors. Signals related to movement execution would then be compared to the ideal movement contained in the motor representation. This view of an 'open-loop' visual control will be developed in Chapter 5.

4. Directional coding of reaching

This chapter deals with integration of mechanisms for direction of movements during reaching at visual objects. One of its basic postulates is that direction of these movements must be coded in a system of co-ordinates (a motor 'map' of space) referred to the body axis, different from the visual map on which the retinal position of objects is specified. Therefore, the internal representation of the visual world in which the subject behaves spatially must include a body reference, if possible in coincidence with objective body position.

The motor and the visual maps may be occasionally superimposed, as when the eyes are aligned with the head and the head with the trunk. In most situations, however, the fact that the eyes move in the head, and the head moves with respect to the trunk means that a single locus in space may correspond to a variety of retinal loci, according to the relative positions of eye, head, and body axes. Reconstruction of the position of objects in a body-centred space should therefore integrate not only the retinal signal documenting the position of the object images on the visual map, but also extraretinal signals related to the position of the eyes in the orbit, the position of the head with respect to the body, and the position of the internal body reference with respect to the body itself. The respective contributions of these extraretinal signals to the construction of body-centred space and to the generation of goal-directed movements are examined in the following sections.

4.1. Eye position signals

4.1.1. Experimental evidence

Evidence for the existence of extraretinal signals derived from eye position comes from experiments showing that goal-directed eye movements are in fact directed at the spatial, not the retinal, locus of the target (see Robinson 1975; Miles and Evarts 1979). Hallet and Lightstone (1976), for example, took advantage of the fact that ocular saccades can accurately reach targets that are presented very briefly, that is, in a condition where they disappear before the saccade is initiated. In their experiment, subjects were instructed to fixate a target that jumped from its position on the central retina to a peripheral position, and then was flashed at another position immediately prior to the saccade intended at the first jump. They found that subjects indeed executed a saccade to the location of the first target,

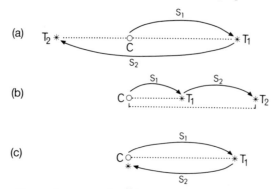

Fig. 4.1 Three illustrative cases of an ideal experiment for demonstrating the spatial rather than retinal, coding of ocular saccades. The basic principle of this experiment is the same as in the Hallett and Lightstone (1976) experiment. In all three cases, C represents the central fixation target, T_1 the first target, T_2 the second target. S_1 represents the extent of the first saccade in response to T_1 and S_2 the second saccade in response to T_2.

In experiment a, target C was turned off as T_1 was turned on, and T_1 was turned off as T_2 was also turned on. Finally, T_2 was also turned off. This complete sequence of events took place before the subject began his first saccade. The saccade S_1 thus corresponded to the retinal signal $C–T_1$. However, the extent of saccade S_2 ($T_1–T_2$) did not correspond to the retinal signal for that saccade ($C–T_2$). Instead, the subject moved to the spatial location of T_2 irrespective of the eye position prior to saccade S_2. The same was true in experiment b. Finally in experiment c, no retinal signal ever existed for saccade S_2, because the location of the second target T_2 was superimposed to that of the central target C.

and from there directed their eyes at the location of the flash. The second saccade was therefore programmed to cover the distance between the location of the first target and the location of the flash, and not the distance between the position of the eyes at the time the second target was flashed and the location of that target (Fig. 4.1). In other words, the second saccade was not programmed according to the amplitude of the retinal signal, but according to the amplitude of the spatial difference between the two targets. In other trials, the position of the flash was on the central retina itself, that is at the same position where the target was before the jump. A second saccade was nevertheless generated in order to bring the eyes back to their initial position, in spite of the fact that, because the eyes were still immobile when the flash was produced, no retinal signal actually ever existed for that saccade. Similar observations were also made by Prablanc *et al.* (1978).

In animals, the dramatic results obtained by Sparks and his colleagues seem to have demonstrated unambiguously the contribution of extraretinal signals to the reconstruction of target position in space. In the experiment of Mays and Sparks (1980) monkeys were trained to look at small targets appearing in an otherwise dark room. The fixation target was extinguished and an eccentric target was illuminated for 100 ms. In this situation the animal normally makes a saccade after about 200 ms. On some of the trials, after the eccentric target was turned off but before the saccade began, the eyes were driven to another position in the orbit by electrical stimulation of the superior colliculus. Nevertheless, the actual saccade was correctly directed at the position of the target in space. This finding supports the hypothesis that an accurate eye position signal is continuously combined with retinal signals to provide a representation of target position in space (see Sparks and Mays 1983). In addition, Guthrie *et al.* (1983) were able to show that the correct final eye position could be reached in spite of intervening perturbations, in animals in which extra-ocular muscle proprioception had been eliminated by bilateral section of the ophtalmic branches of the trigeminal nerves. This finding illustrates the role of feed-forward mechanisms in generating extraretinal signals (see below, Section 4.1.2).

The need for reconstruction of target position in space becomes particularly obvious when one tries to understand the mechanisms for orienting toward targets which stimulate simultaneously the visual and the acoustic modalities. It is known that in many species, the superior colliculus contains superimposed representations of auditory and visual space, such that neurons in the deep collicular layers may respond to visual and auditory stimuli located within the same region of space. This property seems paradoxical: because co-ordinates of auditory space are defined with respect to the ears and hence the head, while co-ordinates of visual space are defined with respect to the retina, the two spaces should become misaligned each time the eyes deviate from the head axis (Poeppel 1973). Studies of eye–head co-ordination in cats, however, have shown that saccadic eye movements are usually followed shortly by head movements in the same direction, so that the eyes rapidly return to the centre of the orbit after each saccade and the co-ordinates of visual and auditory space remain superimposed (Harris *et al.* 1980; Guitton *et al.* 1984). This is not necessarily the case in primates, including humans, where eye position may remain significantly deviated from the head axis following orientation toward a peripheral stimulus (see below). In these species, the position of visual targets has to be reconstructed with respect to the head, so that visual and auditory signals from the same objects are matched. Jay and Sparks (1984) have studied the modalities of this reconstruction at the level of neurons of the intermediate layers of the superior colliculus in awake monkeys. Animals had their head fixed and were in the dark. Auditory targets were presented at a fixed position in space, while gaze position was changed by presenting visual fixation targets at different locations. It was found that neuronal responses to the same auditory stimulus varied as a function of gaze position, indicating that the auditory receptive fields of the neurons shifted with the position of the eyes in the orbit. According to Jay and Sparks

(1984), these findings suggest that 'an internal representation of eye position . . . has been subtracted from the head-centred spatial code of auditory targets . . . '. 'This translation allows the auditory and the visual maps in the intermediate and deep layers of the primate superior colliculus to share a common reference system . . . ' (p. 347).

These facts have been considered in designing models of the saccadic system. Robinson (1975) compared the two possible modes of saccade generation in terms of commands to the saccadic pulse generator. In the retinocentric mode, the command to the pulse generator is: 'go a certain distance in a certain direction'. No account is taken of initial eye position; only the *change* in eye position is important to remove the retinal 'error' between target position on the peripheral retina and central fixation. In the cephalocentric mode, the command is: 'go to a certain location in the orbit', which implies that the initial eye position is taken into account. In order for eye position to influence the pulse generator in the cephalocentric mode, Robinson (1975) postulated the existence of an 'efference copy' (one of the possible modalities for eye position signals; for an account of this concept, see below, Section 4.1.2) derived from the output of the neural integrator, which maintains eye position after the saccade has been completed, and adds to retinal error to create an internal representation of target position with respect to the head (Fig. 4.2). Of course, in the condition where the head is kept stationary, the eye position signal equals zero, and the saccade is driven by retinal error alone, i.e. the system works in the retinocentric mode. A complete account of the problems raised by the transformation of retinal co-ordinates into head co-ordinates can be found in Grossberg and Kuperstein (1986).

4.1.2. The nature of eye position signals

A discussion of the role of eye position signals in reconstruction and representation of target position in space can be traced in the literature as far back as Charles Bell (1823, quoted by Wade 1978). Bell had noticed that visual after-images seem to move with the eye in spite of being stationary on the retina. Hence, Bell thought that ' . . . vision in its extended sense is a compound operation, the idea of position of an object having relation to the activity of the muscles'. 'If we move the eyes by the voluntary muscles . . . we shall have the notion of place or relation raised in the mind.'

Thus, the first step towards understanding how eye movements and/or positions can be monitored (and ultimately consciously perceived) has been to determine the nature of the eye position signals. The hypothesis that has historical priority in this respect (see Grüsser 1984) postulates that a subject can be aware of the 'efforts of will' he directs to his eye muscles, and therefore can use these outflow signals to distinguish between displacement of objects across the retina arising within the external world, from

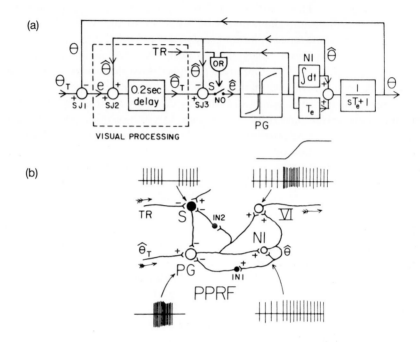

Fig. 4.2 D. A. Robinson's model for the cephalocentric mode of saccade generation.

(a) The retinal error (e) represents the difference between the target position with respect to the head (θ_T) and the eye position with respect to the head (θ). During visual processing of target position an internal representation of the target-re-head position ($\hat{\theta}_T$) is constructed with the help of an efference copy of the eye position ($\hat{\theta}$). It is the comparison between $\hat{\theta}_T$ and $\hat{\theta}$ which generates the signal (e) for the saccade. The saccadic generator (PG = pulse generator, and NI = neural integrator) therefore produces a movement coded in head coordinates. Note that the efferent copy $\hat{\theta}$ of the eye-re-head position is sampled at the output of the neural integrator, i.e. after the steady eye position signal has been generated.

(b) Analogic representation of the model with the known neurophysiological elements (as of 1975) of the brainstem oculomotor plant. The trigger signal (TR) for the saccade inhibits the suppressor unit (S), which releases activity of the phasic unit (PG). As PG is fed with $\hat{\theta}_T$ it generates a burst proportional to the target-re-head position. The tonic neuron (NI) integrates the same activity over time. The output of NI is assumed to be the efference copy $\hat{\theta}$, which is fed back to the phasic cell. Finally the motoneuron (VI) integrates the phasic and the tonic signals to produce the saccade and to maintain the eye in its new position.

Note that empty circles represent excitatory neurons and filled circles, inhibitory neurons. From Robinson 1975, with permission.)

displacement resulting from self-produced eye movements. In its initial formulation (von Helmholtz 1866), this hypothesis represented an explanation of how the visual environment is perceived as stable during eye movements.

A more elaborate theory, though using the same basic notion of a monitoring of the efforts of will, was put forward in 1950. According to this theory, the motor outflow responsible for self-produced movements was thought to be paralleled by neural discharges which represented a 'copy' of the efferent activity sent to the muscles. This 'efference copy' was thought to act on visual centres so as to cancel the interpretation of visual motion inflow to the sensory neurons each time this inflow was a direct product of behaviour (von Holst and Mittelstaedt 1950).

Neuronal responses fulfilling the criterion for an efference-copy type of mechanism have been recorded from visual neurons, first from invertebrate species (Wiermsa and Yamagushi 1967; Palka 1969), and more recently in mammals. In the monkey superior colliculus, for instance, Robinson and Wurtz (1976) found neurons that responded when a visual stimulus rapidly crossed their receptive field, but did not when an eye movement swept their receptive field across the same but stationary stimulus. Neurons in other parts of the central visual system, including the visual cortex, have also been shown to receive eye movement related extraretinal signals [for review, see Jeannerod *et al.* (1979). Also see Vanni-Mercier and Magnin (1982).] These signals might originate either directly from the oculomotor neurons themselves, or from other parts of the brainstem 'oculomotor plant' where neurons with tonic firing rates proportional to eye position can be recorded (see Robinson 1975).

In the behavioural context, the same mechanisms might also account for stabilization of the whole animal, particularly in those animal species whose behaviour is strongly driven by visual motion signals. Experiments reported by Sperry (1943, 1950) seem to demonstrate the effects of efference copy on an animal's behaviour. Sperry (1943) first observed that fish with inverted vision caused by surgical 180° rotation of the eye tend to turn continuously in circles, quite in the same way as a normal fish when stimulated by a visual surround moving at constant velocity. In a later paper Sperry (1950) interpreted this circling behaviour under inverted vision as the result of disharmony between the retinal input generated by movement of the animal and the compensatory mechanism for maintaining the stability of the visual field. Surgical eye rotation, by making the compensatory mechanism in diametric opposition to the retinal input, 'would therefore cause accentuation rather than cancellation of the illusory outside movement'. The mechanism postulated by Sperry was a centrally arising discharge that reached the visual centres as a corollary of any excitation pattern that normally resulted in a movement and was specific for each movement with regard to its direction and extent (see also Teuber 1960).

'Efference copy' and 'corollary discharge' are germane concepts that both imply that outflow information (i.e. arising from efferent systems) can be used at the central level to regulate sensory messages. These two concepts, however, are included in, but do not completely overlap with, the more general concept of extraretinal signals.

Another interpretation of the nature of these signals, in apparent contradiction with the efferent-copy hypothesis, has been offered. Authors such as W. James (1890) held that all incoming messages had to proceed to the brain through sensory channels, and that information flow had to be directed in a sensory-to-motor, rather than in a motor-to-sensory, direction. [For a historical account of this theory, see Jeannerod (1983); see also Chapter 5.] Eye position signals were therefore conceived by these authors as originating from sensory endings within the orbit. This idea was strengthened by the Sherrington (1897) discovery that extrinsic ocular muscles were abundantly supplied with neuromuscular sensory endings and therefore could signal their degree of tension to the nervous system. In addition, Sherrington (1898) had noticed that after the conjunctive and the cornea on both sides were anesthetized, the eyes could still be directed accurately to any given point in a completely dark room. Based on this finding, he deduced that proprioceptive information as to eye position was not conveyed to the brain through the trigeminal (sensory) nerve, but took a retrograde pathway through the motor nerves themselves. Later, however, Sherrington's belief was challenged, and the ophthalmic branch of the trigeminal nerve was shown to be the main pathway for extraocular proprioception (Batini and Buisseret 1974). From there, ocular proprioceptors largely project to the central nervous system. In the cat, responses to eye muscle stretch are found in the cerebellum (Fuchs and Kornhuber 1969; Baker *et al.* 1972; Schwartz and Tomlinson 1977), the superior colliculus (Rose and Abrahams 1975; Donaldson and Long 1980), and the visual cortex (Buisseret and Maffei 1977; Ashton *et al.* 1984).

Experimental data demonstrating the contribution of ocular proprioception in visuomotor behaviour were recently reported in cats. Fiorentini *et al.* (1982) showed that adult cats with unilateral section of the ophthalmic branch of the trigeminal nerve made large systematic errors in jumping from a start box to a luminous target. When the animal was tested with either its deafferented or its normal eye the errors were distributed to the side ipsilateral to the section. This result indicates that unilateral deafferentation may create an unbalance between the proprioceptive projections of the two eyes.

Hein and his colleagues, although they also stressed the proprioceptive contribution to target localization, insisted on the fact that this function was critical during the period of maturation of visuomotor behaviour. Their initial finding was that kittens with surgical eye immobilization fail to acquire visually guided behaviour, as tested with placing responses and visually guided locomotion (Hein *et al.* 1979). Later, Hein and Diamond (1983) were able to directly demonstrate that this effect of eye immobilization was in fact attributable to the absence of proprioceptive volleys from the eye muscles during attempts of the kittens to reach for visual targets. The key experiment was made with kittens first reared in the dark until they were 2–4 weeks of age, then enucleated on one side and maintained in the dark for a few more weeks. At the end of the dark period, the extrinsic ocular muscles of the remaining eye were deafferented by sectioning the ophthalmic branch of the trigeminal nerve. When the kittens were tested for the effects of deafferentation surgery after a few days spent in a normally lighted environment, they were shown to lack visual guidance of their reaching movements. This result therefore suggests that exclusion of proprioceptive input from the eye muscles during development precludes the acquisition of normal visually guided behaviour.

Control experiments (A. Hein, personal communication, June 1986) confirmed that deafferentation must be completed at an early age in order to produce its effect. If, for example, one eye was made blind (by section of the optic nerve), and only muscles of the seeing eye were deafferented, visuomotor behaviour developed normally. Subsequent enucleation of the blind (but proprioceptively normal) eye, had no effect.

In humans, pathological involvement of the ophthalmic branch of the trigeminal nerve (by herpes zoster ophthalmologicus, a disease of viral origin) also produces visuomotor impairment. Campos *et al.* (1986) have studied visuomotor behaviour in six patients with herpes zoster on one side. Five of them made large errors in pointing at targets with their unseen arm, but only when they saw the targets with their affected eye. Pointing errors disappeared on remission of the disease. Because herpes zoster leaves the motor component of the extraocular muscles unaffected, the authors concluded from their observations that the disease mimicked the effects of a section of the ophthalmic branch.

Even though extraocular proprioceptive mechanisms are not yet completely understood [monkey eye muscles have no stretch reflex (Keller and Robinson 1971)], the above experiments demonstrate that proprioception does play a role in encoding eye position. As stressed by Hein and Diamond (1983), 'without inflow from the eye muscles a mobile eye is not localizable in the orbit; without eye movement any proprioceptive input that remains available from the paralyzed eye seems insufficiently informative about eye posture' (p. 132). It remains difficult, however, to sort out the respective contributions of these proprioceptive signals and of signals related to corollary discharge mechanisms for constructing a stable spatial representation and for controlling visuomotor behaviour.

4.1.3. Role of eye position signals

In humans a number of experiments have been designed for testing the role of eye position signals in the localization of targets within extrapersonal space. These experiments have often dealt separately with the various aspects of the problem, such as perceptual stability of the environment during eye movements, subjective determination of spatial co-ordinates, and visually goal-directed movements. It will be shown, however, that these aspects are not separate entities, but rather are components of the same functional mechanism, that can be dissociated by the experimental conditions in which the subjects are tested.

(a) Perceptual stability

Historically, the notion of eye position signals is not only associated with the problem of building up a reference for goal-directed movements (which is the primary concern of the present chapter), it is also involved in the explanation of visual 'stabilization' during eye movements (see von Helmholtz 1866).

   In fact, a series of experiments by the Matin group has shown eye-movement related extraretinal signals to be inappropriate to account by themselves for perceptual stability during eye movements. Matin and Pearce (1965) required subjects to localize with respect to each other two sources of light briefly presented in close succession, in association with an ocular saccade. The subjects, who had to report whether the second light appeared to the right or left of the first, eventually made large spatial errors. These errors occurred not only when the stimuli were presented during the saccade itself but, in fact, during a span of time extending from about 200 ms prior to the saccade up to several hundred ms after its completion. In addition, spatial distribution of the errors suggested that the perceived space was contracted in the direction of the eye movement (Matin and Pearce 1965; Matin 1972). These results therefore indicate that the saccade-related extraretinal signals would be too weak to compensate for the perceptual effects of eye movements (subjects made localization errors), and not properly timed to the saccade. If, as argued by Matin (1972), perceptual stability is nevertheless achieved in the conditions of everyday life, this is because other mechanisms add to the extraretinal signal. Purely visual mechanisms, like those related to the retinal slip occurring during the eye movement, would also intervene for masking undesirable visual effects of saccades (Campbell and Wurtz 1978; O'Regan 1984).

   The localization errors observed by Matin and Pearce, however, might be related to the way the subjects processed spatial information during eye movements, rather than to the inadequacy of spatial localization mechanisms. Indeed, if subjects in the same experimental situation as that of

Matin and Pearce, used a different mode of spatial processing, i.e. if they attempted to achieve visuomotor localization rather than to report their perceptual experience, they became very accurate. This fact was clearly demonstrated by Skavenski and Hansen (1978) when subjects were required to strike with a hammer at the location of test lights presented for a brief period of time on the trajectory of ocular saccades. Despite their impression of making large errors, the subjects hit the targets with remarkable accuracy. This result shows that spatial invariance is preserved behaviourally, if not subjectively, during eye movements, and that eye movement-related extraretinal signals are involved in generating relevant information for this invariance.

(b) Lack of eye 'position sense'

Position sense usually refers to the contribution of afferent mechanisms to subjective experience about respective positions of limb segments (but see Chapter 5). In the case of eye position with respect to the head, experimenters have failed to demonstrate directly the existence of such sensations. Brindley and Merton (1960) reported the perceptual consequence of forced duction of the eye with a forceps (with an anaesthetized cornea and in the absence of vision with that eye). Passive rotation of the eyeball of 20° or more was undetected by the subject. When the eye was maintained immobile with the forceps, attempts by the subject to move it were accompanied by a feeling of movement. These results merely indicate lack of contribution of neuromuscular spindles to extraocular 'muscle sense', although they do not exclude the possibility that other types of mechanoreceptors, also blocked by anaesthesia of the cornea and the conjunctiva, might normally contribute to 'position sense'. Other experiments, however, demonstrate that this is not the case. Jeannerod *et al.* (1965) recorded eye movements in subjects attempting to look in the dark at previously learned target positions. Subjects were unable to reproduce these target positions: while their saccades were directed in the proper direction they were of an exaggerated amplitude (Fig. 4.3(a)). This tendency to overshooting learned target positions was further increased when the attempts were made with the eyes closed. Poor performance of subjects in reproducing target positions contrasted with their impression of being quite accurate (Jeannerod *et al.* 1965; see also Lenox *et al.* 1970, Koerner 1975a). More recently, Allik *et al.* (1981), using an electromagnetic eye movement recording technique (more accurate than the EOG technique used by previous investigators), showed that subjects were unable to reproduce simple eye movement trajectories behind closed eyelids. Rather, they tended to make highly hypermetric movements in poor topological correspondence to the previously observed visual pattern (Fig. 4.3(b)).

Position sense should also be involved in judging the direction of a point source of light in the dark. In the absence of a visual frame of reference the

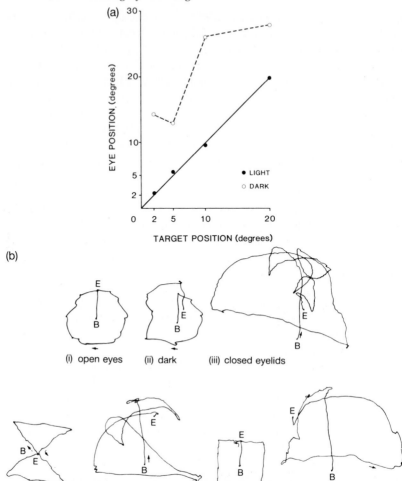

Fig. 4.3   Defective duplication of eye positions in the absence of vision.

(a) Subjects first track a visual target jumping from a central position to prede-termined positions at 2, 5, 10, and 20 ° in the horizontal meridian. Eye position, recorded by the electro-oculographic technique, appears to be linearly related to target position (filled circles, continuous line). Subjects then attempt to duplicate the same positions by moving their eyes in the dark. In doing so, they largely over-shoot the learned target positions (open circles, dashed line). Results from 14 sub-jects. Replotted from Jeannerod *et al.* (1965).

(b) Attempts at duplicating simple eye movement patterns. Upper row: (i) the subject first tracks small target lights arranged in a circle (10° radius). Then, he tries to duplicate the same trajectory with the eyes open in the dark, (ii) and with closed lids (iii). Lower row: same experiment with another subject. Each shape is first examined and then duplicated behind closed eyelids (B = beginning; E = end of eye trajectories). Eye movements recorded by electromagnetic technique. Note poor duplication of eye positions and general tendency to overshooting. (From Allik *et al.* 1981, with permission).

only way to determine the position of a target in space is to infer it from its position on the retina with respect to the eye axis, and from the position of the eye axis with respect to the head–body axis. Lack or paucity of information about eye position should therefore result in inaccurate judgements. This was confirmed in experiments where subjects were required to indicate the position of a briefly flashed target with respect to another, previously fixated, target. Matin and Kibler (1966) had their subjects monocularly fixating a small light for 4 s in an otherwise dark room. The fixation period was followed by a 3 s period of total darkness during which the subjects attempted to maintain the same fixation position. This dark interval was followed by a 100 ms flash appearing either right or left of the previous fixation target. The subjects had to decide whether the flash appeared at the same position as the fixation target, or right or left to it. If the fixation target corresponded to a primary gaze position, subjects' judgements on flashed target direction with respect to the eye axis were accurate to the nearest 20 min of arc. If the fixation target was far from primary position (e.g. 35°), errors as large as one degree were made by the subjects. Matin and Kibler's conclusion was that accuracy in judging visual direction was relatively poor, especially if the judgement had to be made with respect to a position of gaze different from primary position. Their interpretation for such inaccuracy was that subjects were in fact unable to maintain their eyes in the previous fixation position during the dark interval. The same authors (Matin *et al.* 1966) were able to show by direct measurement of eye position that involuntary drifts of as much as one degree or more occurred during the 3 s dark interval.

The Matin and Kibler interpretation was confirmed by Fiorentini and Ercoles (1966), who showed that visual direction of the source could be reported with a small variance relative to its retinal location, but that reports differed systematically from the expected values (those relative to the actual eye position) by an amount that increased with the duration of the dark interval. This result indicates that the spatial position of a target can be accurately detected based on its retinal 'local sign', but that unnoticed and uncorrected drifting of the eye in the dark alters the relationship between the local sign of the target and its actual position with respect to eye axis.

There are disagreements in the literature about the amplitude and direction of involuntary eye drifts in the dark. Skavenski and Steinman (1970) showed that the eyes could be kept within 2° of target position for more than 2 min after the target was turned off. Skavenski (1971) recorded eye position for longer durations during attempts at maintaining fixation in the dark. In two subjects the eye position drifted progressively by 3–4° after 7.5 min. In addition, small saccades in the corrective direction were recorded during these attempts at maintaining fixation. Skavenski (1971) concluded from these results that eye position could be maintained relatively well in the dark. It must be reminded, however, that other authors reached an opposite conclusion. Allik *et al.* (1981), using their magnetic coil recording technique showed that either in darkness or with lid closure the eyes drifted in all directions by as much as 5–10° around the required position during attempts to maintain fixation for 1 or 2 min. These different results might be explained by different degrees of training of the subjects.

Finally, another function where it seems reasonable to assume that eye position signals should normally contribute is 'direction constancy', whereby a stationary object is perceived at the same location in spite of changing eye position. Direction constancy holds in a natural, lighted, environment. In the dark, however, small luminous objects are commonly perceived to jump during saccadic eye movements, usually in the direction opposite to that of the movements. This observation demonstrates incomplete direction constancy in the dark. Studies by Hill (1972) and LaVerne Morgan (1978) have confirmed a systematic underconstancy in the perception of visual direction in the dark. In the LaVerne Morgan study, subjects (with the head fixed) had to determine the position of test flashes with respect to their subjective midline, while fixating stationary targets at different locations in their visual field. Results of this experiment showed systematic departure from constancy of direction of the test-flash, which was perceived shifted in the direction of the eye fixation. The amount of the shift was proportional to the amplitude of the eye deviation from its primary position. LaVerne Morgan's conclusion was that the systematic error in perception of direction of the test-flash was due, at least in part, to misregistration of eye position.

Taken together, the above results suggest relatively poor position sense in human eyes. Because the neurophysiological apparatus for detecting eye movements and/or positions is clearly present, poor position sense could be due to absence of reliable encoding of eye position signals at the central level. It is likely that the signals generated either by the motor commands or by the muscles themselves must be calibrated by vision before they can become truly informative. When retinal feedback is available, eye position remains stationary during gaze fixation, perceptual stability is achieved during saccades, and direction constancy is preserved during gaze displacements. But when retinal feedback is absent or degraded, eye position signals alone prove insufficient for maintaining perceptual invariance of the environment.

4.1.4. The 'paralysed-eye' situation

The situation where the eyes are pathologically or experimentally paralysed has long been considered an ideal paradigm for demonstrating the monitoring of the eye's position in its orbit by the central nervous system.

As first described by von Graefe (1870), people with paralysis of an extrinsic eye muscle display striking behaviour when they attempt to reach
toward objects in their peripheral visual field viewed only with their paralysed eye. Typically, they overreach in the direction of the attempted eye
movement, which is prevented by the paralysis, and miss the target ('past-
pointing'). Attempts to move the eyes against the paralysis may produce an
illusory displacement of the visual scene in the direction of the attempted
movement. These phenomena have been documented by clinical cases
(Jackson and Paton 1909; Adler 1943; Perenin *et al.* 1977; Von Norden *et
al.* 1971), and by experiments that used reversible block of extraocular
muscles in normal subjects (Kornmüller 1931; Siebeck 1954; Brindley *et al.*
1976; Stevens *et al.* 1976; Matin *el al.* 1983).

(a) Pathological observations

Perenin *et al.* (1977) studied four patients with a complete paralysis of
either the sixth or the third nerve serving one eye. Patients were tested for
pointing by hand at visual targets appearing in their peripheral visual field.
The normal eye was occluded, the head was fixed, and sight of the arm
used for pointing was prevented. In addition, movements of the occluded
normal eye were recorded by the EOG technique. When targets were presented to the affected eye in an area of the visual field that corresponded to
non-paralysed muscles, the patient pointed correctly at the target location.
By contrast, when targets appeared in the area of the visual field corresponding to the paralysis, the hand was directed far more distal to the midline than the actual location of the targets. For example, target
presentation corresponding to a retinal locus 30° from the fovea could yield
a pointing movement directed 50° from the midline. This constant error, a
typical example of past-pointing, is shown in Fig. 4.4. Movements of the
normal (covered) eye directed toward targets presented to the paralysed
eye were also of an exaggerated amplitude, in such a way that they would
have clearly overreached target position (Fig. 4.4(b)). Von Graefe's interpretation of this phenomenon was that an effort larger than normally
required for a given result had to be produced and that the increased sensation of effort yielded an overestimation of the rotation angle made by the
eye, together with the illusion of the visual field being displaced in the same
direction. According to this interpretation, past-pointing would be a direct
consequence of the estimated eye position. An observation made by Perenin *et al.* (1977) seems to confirm the von Graefe view. They asked one of
their subjects with extra-ocular paralysis to keep his gaze fixated at the
midline during presentation of the targets. His pointing movements were
correctly directed at the location of the targets (Fig. 4.4(a) left).

The notion of an exaggerated effort exerted against the paralysed muscle
postulated by von Graefe for explaining past-pointing belongs to the same
theory as von Helmholtz's notion of efforts of will. Another plausible

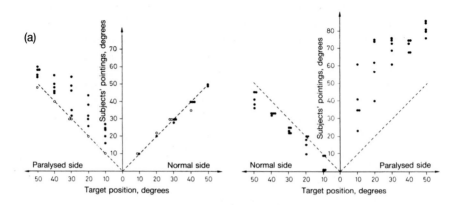

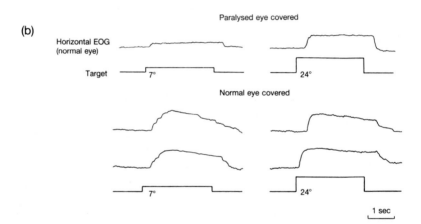

Fig. 4.4 Relationship of hand pointing errors to exaggerated oculomotor output in subjects with paralysis of extrinsic ocular muscles.

(a) Past-pointing is clearly observed in two patients with a paralysis of the left external rectus, and of the left internal rectus, respectively. In the patient displayed on the right side of the figure, filled circles represent pointing trials with free eye movements. Open circles represent trials where the subject has been required not to move the eyes.

(b) Eye movement recordings from the *normal eye* during pointing trials in a patient with a paralysis of the left internal rectus [same patient as in (a), left part of the figure]. Upper row: the patient sees the targets with his normal eye (paralysed eye is covered). Note normal saccadic eye movements. Lower row: the patient sees the targets with his paralysed eye (normal eye is covered). When the targets appear to the right (upward deflections) the patient largely overshoots target positions with his covered eye. Two trials are represented for each target position. (From Perenin *et al.* 1977.)

explanation for the increased oculomotor output during attempts to move against the paralysis is that the paralysed eye cannot foveate the target. In that situation, the error between the retinal position of the target and the actual position of the eye cannot be cancelled by a saccade, and the oculomotor system behaves as if it were constantly fed by the same error signal. A similar effect can be produced experimentally by feeding the output of the oculomotor system into target position with a positive feedback: the subject produces iterative saccades toward the target without being able to catch it (Young and Stark 1963). This explanation preserves the possibility of an exaggerated oculomotor output which, if monitored centrally, can account for past-pointing with the hand. It could be argued that, because eye and arm movements tend to be coupled in the action of pointing (Biguer *et al.* 1982; see Chapter 2), parametrization of the motor commands occurs over the total coupled system. As a consequence, the increase in force required to move the paralysed eye is also distributed to the system controlling the arm (Kelso *et al.* 1981). In the case where the subject fixates straight ahead while reaching for the target (as in the example shown in Fig. 4.4(a)), thereby decoupling his eye and arm movements, past-pointing disappears.

(b) Experiments in normal subjects

Results from other experiments in normal subjects confirm the latter explanation for past-pointing. In these experiments (Skavenski 1972), subjects were required to maintain monocular fixation of a point source of light in the dark, while their eye was constantly loaded by mechanical traction. This situation of load involved neither change in retinal locus of the target image nor eye movement, since fixation was maintained actively. The only variable was the force exerted by the subjects to oppose ocular displacement by the traction. Subjects were requested to indicate the perceived direction of the fixation target by placing a second, movable, target in their subjective 'straight ahead' position. The fixation target was perceived as being displaced contralaterally to the direction of the load, and the amplitude of this displacement was roughly proportional to the load. These results support the conclusion (Skavenski 1976) that perceived target direction is determined by the magnitude of outflow signals related to eye position, and that this information is directly involved in subjective determination of spatial co-ordinates, therefore supporting the classical notion of a monitoring of the efforts of will.

The effects, on spatial localization, of experimental paralysis of eye muscles in normal subjects, lead to the same general conclusion. In the experiments reported by Matin *el al.* (1983), the subject was partially paralysed by systemic injections of d-tubocurarine, which limited the range of ocular fixations that could be maintained. Within this limitation, however, maintenance of fixation in any given position was not accompanied by a greater sensation of effort than it is in the normal observer. During experiments the subjects were requested to fixate a target appearing in front of them at horizontal eye level. When all lights except for the fixation target were extinguished, the target appeared to move slowly downward. With re-illumination of the room the target seemed to return immediately to its original height. Both magnitude and direction of the illusory target movement in the dark were dependent on the position of the head. When the head was tilted backward with respect to the vertical, the target appeared to move downward. Decreasing the head tilt angle with respect to the vertical reduced the apparent movement of the target. If the head was tilted forward instead of backward, the target appeared to rise. Systematic variation of target height with respect to eye level revealed that illusory displacement was a function of the eye's position in the orbit, and not of head position with respect to the vertical. The more the target position deviated from eye level, the larger was the illusory displacement in the dark. When target position corresponded exactly to the position of the gaze axis in space, no apparent displacement was noticed by the subjects.

These somewhat extreme experiments seem to provide a decisive clarification as to the role of eye position signals in determining spatial localization of targets. In the situation described by Matin *el al.* (1983), the tilt of the head backward or forward with respect to the vertical induced neural commands to the corresponding eye muscles to maintain the gaze in its horizontal position and to achieve target fixation (i.e., to muscles lowering the eye in case of backward tilting, and to muscles elevating the eye in the case of forward tilting). Because of the paralysis these neural commands were inefficient in bringing the eyes to the required position, but the signals they provided were nevertheless monitored by neural structures involved in space perception. In the absence of a visual frame of reference (e.g. in darkness), these signals were used as an index of the gaze direction and, consequently, of the position of the target with respect to the body. Discrepancy between command signals and actual gaze position accounted for apparent movement and mislocalization.

It has been shown that past-pointing and illusory displacement of targets occur only when the eyes are partially paralysed (as in the Matin *et al.* 1983 experiments), and not in the situation of total paralysis (Brindley *et al.* 1976). This confirms the idea that the critical information for producing these effects may not be the extraretinal signal alone, but rather the discrepancy between the retinal and the extraretinal signals. When the two are correctly matched, space remains invariant, although mismatch between the two produces instability and mislocalization.

4.1.5. Experimental deviation of gaze posture

The spatial consequences of ocular misalignment, whether the result of pathology or surgery, also provide arguments for understanding the contribution of eye position signals to visuomotor localization. Mann *et al.* (1979) examined eye–hand co-ordination in humans with monocular strabismus that had appeared early in childhood. It is known that such subjects seem to use only their visually aligned eye for localizing objects. Vision with the deviated eye is constantly suppressed whenever both eyes are open. Mann *et al.* (1979), using tests which required the subjects to point a hand toward targets viewed by the deviated eye, demonstrated constant errors, the direction of which reflected the momentary orbital position of the visually aligned, 'dominant', eye. They proposed that during visuomotor development these subjects had used only their dominant eye in constructing their spatial reference frame and that position signals from that eye had become critical for localizing objects.

Other experiments in animals and humans tend to confirm that a stable reference mechanism is established during development on the basis of eye position signals. This reference is subsequently used for visual guidance and is not modified by later alterations in eye position. Olson (1980) examined in cats the effects of visual localization of surgically induced monocular strabismus. He measured the accuracy of jumps onto a narrow platform. Jumps guided by the normal eye were accurate, those guided by the eye that had been operated on were systematically displaced in the direction opposite to the eye deviation. Olson interpreted this finding by suggesting that the cats were unaware of the eye deviation and continued to use a 'registered' eye position in conformity with their head axis (as cats normally do, see above), to guide their visuomotor behaviour. Therefore, in jumping in the direction of their head axis, they systematically landed in the direction opposite to the target. Steinbach and Smith (1981) reported similar results in humans operated on for strabismus. These patients were tested shortly after surgery for hand pointing accuracy under control of the eye that had been operated on. A large systematic error was observed in some of them. The direction of the error was consistent with the Olson hypothesis that the brain was unaware of the surgical eye rotation and still relied on eye position information corresponding to the pre-surgical eye position to direct the pointings. This effect, however, was observed only in patients who had been operated on several times, and in whom the extra-ocular muscle tendons had been damaged. In patients operated on for the first time, pointings were not shifted because, according to the Steinbach and Smith interpretation, tendon organs were still intact and provided the central nervous system with afferent signals corresponding to the true eye position. This interpretation therefore implies that eye position signals can originate from both inflow and outflow sources. Outflow information might be used as the short-term indicator of eye position, whereas inflow information would allow for modification and recalibration of this fast system over longer periods of time (Steinbach and Smith 1981).

4.1.6. Role of eye position signals in pointing accuracy

The above experiments have only provided indirect demonstration for the contribution of eye position signals to directional coding and accuracy of goal-directed movements. Other studies have attempted to test this contribution more directly. These studies, however, have remained mostly inconclusive.

Prablanc *et al.* (1979a), using the same experimental situation as described in Chapter 3, measured pointing accuracy in 'visual-feedback' and 'no visual feedback' conditions, with the additional constraint that eye movements were not allowed during the pointing task. Subjects were instructed to keep their gaze fixed on the midline target while they pointed at other targets appearing in their peripheral visual field. In the 'visual feedback' condition, pointing accuracy was little affected (the variable error was slightly increased) by the absence of eye movement toward the targets at which the hand pointed. In the 'no visual feedback' condition, pointing accuracy was severely degraded, though not to a significantly greater extent than in the 'no visual feedback' condition with eye movements allowed.

This result therefore does not seem to favour a role of eye position signals in visuomotor control. However, other experiments also using pointing movements as an index, suggest different conclusions. In the Festinger and Canon (1965) study, target lights were presented at different positions within the visual field. In each trial subjects had to point in the dark at the position of the target after it had disappeared. If the target jumped rapidly from the central position to one of the peripheral positions, pointing was relatively accurate. If, however, the target moved slowly from central position to its final position before disappearing, large pointing errors were made. Interpretation of this result was based on the difference between mechanisms generating saccadic eye movements (which were present in the target jump condition) and smooth pursuit eye movements (which were present in the other condition). Saccadic eye movements are directed at stimuli located on the retina outside the fovea, while smooth pursuit relies on velocity signals across the fovea. Festinger and Canon (1965) concluded from their results that efferent command signals related to target position yield to better visuomotor localization than signals related to target velocity. Another study by Honda (1984) also involved manipulation of the saccadic signal putatively used for target localization. Subjects had to point with their unseen hand at the location of a visual target after it had disappeared. Before stopping at its final position and disappearing, the target jumped to one or several other positions, at such a rate that a saccade was generated for each jump. The results showed that the pointing error increased with the number of jumps that occurred before the target stopped [compare this result with that of Bock and Eckmiller (1986), showing an increase in pointing error when target is displaced at successive locations]. Honda's conclusion was that subjects were not able to use the cumulative eye position signal generated by addition of successive saccades, but rather tended to use the signal corresponding to the largest saccade in the sequence.

Experiments such as those of Festinger and Canon (1965) and Honda (1984), where oculomotor signals related to target position in space were directly manipulated, give more positive indications concerning the role of these signals in visuomotor behaviour. However none of these experimental situations, which involve head fixation with respect to the body and where only the eyes move, correspond very closely to real life conditions. It seems more reasonable to assume that in the normal situation, where the head is free to move, the relevant signals are derived from both eye and head positions and contribute to a *gaze position* signal. This possibility will be examined in the next section.

4.2. The contribution of head position signals

At variance with the relatively poor signalling of eye position in orbit, head position with respect to the trunk seems to play an important role in encoding target position in space. In the monkey, Cohen (1961) showed that abolishing proprioceptive information from the neck muscles (by sectioning the dorsal roots) rendered the animals unable to reach accurately for objects, even when their eyes were fixating these objects. In humans, Marteniuk (1978) provided evidence that orienting the head toward a target consistently facilitated accurate localization of that target. In addition, the same investigator has shown that head position could be used by subjects as a landmark for reproducing the position of a target previously fixated by a combined movement of head and eyes.

4.2.1. The effects of vibrating the neck muscles

Contribution of head position signals to visuomotor behaviour seems to be best demonstrated by a simple experiment reported by Biguer *et al.* (1986). These authors took advantage of the distortion of position sense produced by muscle vibration. It has been known since Goodwin *et al.* (1971, 1972) that vibrating a muscle can induce an illusory movement of the corresponding segment. If the vibrated limb is prevented from moving, it is nevertheless felt to move. In their experiment, Goodwin *et al.* vibrated the biceps brachialis in one arm, in blindfolded subjects. The vibrated arm was immobilized, the other arm was used to 'track' the illusory movement. With their tracking arm subjects consistently indicated that they felt their vibrated arm more extended than it actually was. The explanation to this finding was that, the Ia discharges set up by the vibration are interpreted by the sensorium as due to a stretch of the biceps muscle, and thus taken to indicate that the joint is more fully extended that it actually is. This might perhaps be through some higher centre recognizing a mismatch between the actual state of the muscle and that which was 'intended' by the controlling centres. (Goodwin *et al.* 1971).

Biguer *et al.* applied vibrations at 100 Hz to the posterior neck muscles on one side. Their subjects were asked to maintain visual fixation on a dim luminous target which appeared directly ahead of them in an otherwise dark room. When vibration was applied, subjects reported apparent displacement of the fixation light. This illusory displacement was usually in the horizontal dimension and to the side opposite to that of the stimulation but, by altering the exact location of the vibrator, illusions of vertical or diagonal movement could be produced. The target initially showed both motion and displacement. Displacement ceased within a second or two. Afterwards the target appeared to continue to move without further change in position. Apparent movement persisted as long as vibratory stimulation was maintained. When vibration was discontinued, the light appeared to move in the reverse direction for a brief period of time.

Control experiments showed that this effect was not due to either eye or head movements during vibration. Indeed, the target motion was still perceived during vibration by subjects whose head was immobilized with the aid of a bite-plate. In addition, no eye movements could be detected on EOG recordings. Finally, in order to eliminate the possible role of very small eye movements (i.e. beyond the scope of the EOG technique), an opaque mask was placed in front of the subject's eye. A 0.5 mm vertical slit was made in the mask at the level of the subject's gaze axis, so that he could see the target. When vibration was applied to the neck at the proper location, the subject still experienced the illusion of motion of the target in the horizontal dimension. If the eye had moved during vibration, the subject would unavoidably have lost view of the target.

In accord with the explanation of Goodwin *et al.* vibration of the neck muscles on one side (e.g. the left) produced the same afferent spindle discharge as if the neck muscles were stretched by head rotation toward the opposite side (to the right in this example). This illusory change in head position was interpreted as apparent displacement of visual objects to the right. Therefore, if a subject were asked to point at a visual object under this condition of vibration, his pointing should err in the direction of the apparent displacement. This is exactly what Biguer *et al.* reported. When their subjects had their left neck muscles stimulated, they systematically pointed to the right of the actual target position. Figure 4.5 shows the mean location of pointings before, and during, vibration of the neck muscles on the left side. Vibration induced a shift of the trajectory of pointings to the right in all subjects. The loci of pointings before and during vibration did not overlap in five of the six subjects tested. For these five subjects, the

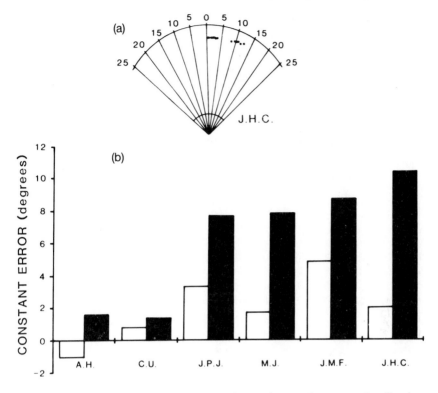

Fig. 4.5 Effects of vibrating posterior neck muscles on visuomotor localization. Accuracy of pointing with right hand at a visual target placed at midline was measured in six head-fixed subjects. Most subjects showed a constant error in pointing too far to the right [part (b), empty bars]. During vibration (100 Hz) of the left posterior neck muscles, pointing was deviated to the right (black bars). This effect was significant in all subjects except C.U. (but see Fig. 4.6).

Part (a) represents raw data from subject JHC. Note pointing positions clustered around 2° from midline before vibration and around 10° during vibration. (From Biguer *et al.* 1986.)

mean displacement of the pointings was 5.0° (SD, 2.1°). The sixth subject reported that the visual illusion was present only for a short time after the vibration began. In fact, a detailed look at the individual pointing movements performed by that subject revealed that the first pointings were deviated to the right, as expected, but that this effect disappeared during the course of the stimulation (Fig. 4.6). Displacement of the pointings was thus correlated with the intensity of the visual illusion.

This experiment not only confirms the contribution of spindle afferents to position sense in general, it also demonstrates clearly that perturbations in head position sense are directly translated into mislocation of objects in space and misdirection of visuomotor behaviour.

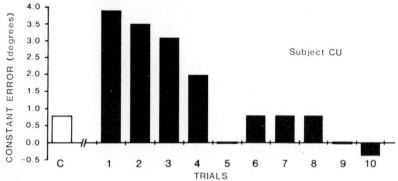

Fig. 4.6 Relationship of pointing bias during neck muscle vibration and illusory visual displacement.

Pointing performance of one subject (CU) involved in the experiment described in Fig. 4.5 was examined in detail. This subject mentioned at the end of the vibration session that he had lost the illusion of displacement of the visual target during the course of stimulation. His first three or four pointings were strongly deviated to the right, although this effect suddenly disappeared at a later stage. This change in behaviour was likely to have corresponded to the loss of visual illusion as mentioned by the subject. (From Biguer *et al.* 1986.)

4.2.2. Influence of head mobility on pointing accuracy

The precise role of head, or conjugatged head–eye position signals in the encoding of target location was examined in an experiment where subjects were tested for pointing accuracy, with their head either fixed or free to move. The latency, duration, and accuracy of eye, head, and hand movements were measured. Gaze position in space (i.e. the sum of eye and head positions) was reconstructed electronically. Subjects were instructed to point as quickly and accurately as possible at randomly located peripheral targets. Movements always started from the midline position. Experiments were performed in the dark, so that no visual reafference form the hand movement was available at any time (the previously described 'no visual feedback' condition). The target remained visible throughout the duration of each trial. Two experimental sessions were run for each subject: one in which the head was free to move ('head-free' condition), and one in which the head was fixed in the straight ahead position ('head-fixed' condition) (Biguer *et al.* 1984) (Fig. 4.7).

The latency data from this experiment confirmed those obtained by Biguer *et al.* (1982), i.e. eyes moved first, followed by the head (in the head-free condition), and then the hand (see Chapter 2). In addition, eye and hand latencies were not affected by whether the head was fixed or free to move. Durations of eye, head, and hand movements were significantly affected by target eccentricity, tending to increase with distance from midline. Experimental conditions (e.g. 'head-fixed' or 'head-free') did not

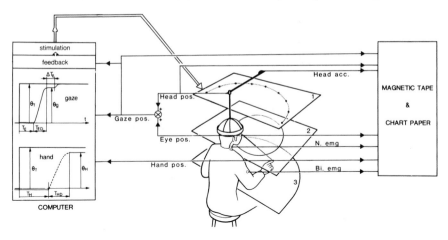

Fig. 4.7 Experimental display for studying eye–head–hand co-ordination. The arrangement is basically the same as that described for studying visual control of pointing (see Fig. 2.9). In this apparatus, targets are presented on a circle centred at subjects body axis. Vision of the hand can be allowed or excluded. Head movements and positions can be recorded through a helmet fitted to the subject's head. The helmet is connected with an accelerometer for recording head movement onset. It is also connected with a potentiometer for recording head position. The output of the potentiometer is combined with the output of eye position recording, for reconstructing the gaze position. Finally EMG activity of the right neck muscles and muscles of the right arm can be recorded. (From Biguer *et al.* 1984.)

affect these durations. Latencies and durations of movements were such that head movement was always completed before hand movement, i.e. the time interval between the end of the head movement and that of the hand movement (TIHH) was always positive. The mean TIHH ranged between 110 ms for targets located at 10° from the midline and 160 ms for targets located at 40° (Fig. 4.8). Similarly, the time interval between the end of the gaze movement and the end of the hand movement (TIGH) was also largely positive. It ranged between 243 ms and 375 ms on average for targets located at 10° and 40°, respectively (Fig. 4.8).

Pointing accuracy of hand movements was strongly affected by whether the head was fixed or free to move. In the 'head-fixed' condition, large constant and variable errors were observed. Both types of error tended to increase as a function of target distance from the midline (Fig. 4.9). The fact that in the present experiment constant errors were in the direction of overshooting instead of undershooting, as is usually observed in visual 'open-loop' pointing, should be attributed to a difference in the type of movement made by the subjects. Pointing experiments usually involve targets located at arm length, i.e. arm movements involve mainly rotation at the shoulder joint. By contrast, in the present experiment, because targets

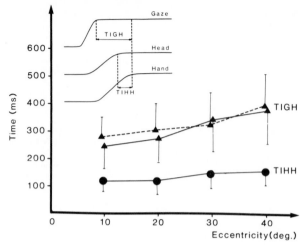

Fig. 4.8 Temporal organization of the action of reaching with the head fixed and the head free.

Time interval between the end of gaze movement and the end of head movement (TIGH, triangles) has been represented as a function of target eccentricity, in the head-free condition (solid line) and the head-fixed condition (broken line). Dark circles = time interval between the end of the head movement and the end of hand movement (TIHH) as a function of target eccentricity, in the head-free condition. Insert on the left indicates how TIGH and TIHH were computed. (From Biguer *et al.* 1984.)

were presented on a circle around the body axis at a distance of about 40 cm (Fig. 4.7), pointing movements involved simultaneous rotation of the shoulder and flexion of the elbow. It may be that this type of biarticular movement is calibrated differently.

In the 'head-free' condition both constant and variable errors were consistently reduced with respect to the 'head-fixed' condition. In addition, the usual relationship of error size to target distance was no longer observed (Fig. 4.9). Statistical analysis of errors in the two conditions (Biguer *et al.* 1984) showed that improvement in pointing accuracy in the 'head-free' condition was significant mainly for targets located far from the midline (40°).

Similar results were obtained by Roll *et al.* (1981, 1986) in an experiment involving an aiming task. Subjects with their head fixed or free had to direct a hand-held spot source of light toward targets. The position of the spot with respect to the target was recorded photographically. With their head fixed, the overall group of subjects were found to undershoot targets. This constant error increased with eccentricity of targets, together with the variable error. When the head was free to move, the constant error disappeared, and the variable error was decreased.

There are a number of possible explanations for the improvement in

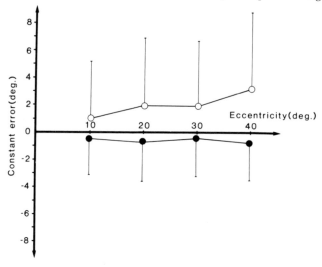

Fig. 4.9 Influence of the head-free condition on pointing accuracy. Subjects pointed at luminous targets appearing in the dark, i.e. without possibility to see the moving limb. When the head was fixed, large constant and variable errors were observed (open circles). Constant errors were in the direction of overshooting target positions, which can be explained by the disposition of targets on a circle around subjects body axes (see Fig. 4.7). In the head-free condition, the constant errors disappeared and the variable errors were greatly reduced. (From Biguer *et al.* 1984.)

pointing accuracy in the condition where co-ordinated eye and head movement is allowed. First, the synergy of eyes, head, and hand moving together may represent an advantage for terminal accuracy. These segmental movements during the action of reaching belong to the same motor ensemble, their commands are released synchronously and they are linked by a relatively strict temporal co-ordination. Artificial disruption of this ensemble, by head blocking for example, could introduce a mechanical limitation of its performance.

Alternatively, proprioceptive signals generated by the head movement itself (e.g. originating from neck joints, tendons, or muscles) may facilitate the programming of target-directed hand movements and improve their directional coding. The interval between the end of the head movement and the end of the hand movement (as measured by the TIHH, see above) is largely compatible with the time required for kinaesthetic afferents to influence the course of a movement (see Chapter 5). However, it is likely that, in the 'head-fixed' condition, the neck kinaesthetic receptors adapt over time so that the signal they normally provide about head position degrades rapidly. Paillard and Brouchon (1974) have suggested in a different context that the position of a limb can be accurately perceived only for a short time after it has been moved actively by the subject, but not when the same position has been maintained for a period of time exceeding 15 s (but see also Chapter 5).

A third explanation for the improved hand accuracy in the 'head-free' condition is based on a hypothesis concerning the factors of the pointing error. It may be assumed that the pointing error integrates not only some motor variability, but also errors inherent in the encoding of eye and head position. The error in encoding eye position may remain relatively low for small angular displacements of the eye within the orbit. Beyond a certain limit, however, the error may increase sharply. Head movements would thus occur to maintain the eye position within the limit where it can be accurately encoded. This hypothesis would be consistent with the fact that, in most subjects, head movements occur when they orient toward targets farther than about 10°. This value might correspond to the limiting angular displacement of the eye, beyond which eye position signals are degraded. The same argument holds for the error in estimating the head position with respect to the body—that is, beyond a certain angular displacement of the head, the head position signals would become less accurate, and rotation of the trunk would have to occur. This hypothesis is formalized graphically in Fig. 4.10.

4.3. Signals related to body axes

This section explores the possibility that there are mechanisms that encode the position of body axes. It is postulated that visual, somatosensory, vestibular (and possibly other) signals may generate an internal representation of 'egocentric' co-ordinates; and that these represented co-ordinates provide a reference for actions within personal space (like touching a part of one's own body), as well as for actions directed at objects within extrapersonal space.

Several concepts have been used to account for the fact that egocentric body co-ordinates can be perceived subjectively. Schilder (1935), for instance, included the body axis within the more general framework of the 'body scheme'. Similarly, Werner and Wapner (1952) postulated that perception of spatial co-ordinates reflected the innervation pattern of the whole organism during acquisition of targets within proximal space (the sensory-tonic field theory, see below). Observations of orienting and locomotory behaviour following unilateral lesions of the nervous system have provided experimental bases to these hypotheses.

4.3.1. Representation of body co-ordinates in normal subjects

The ability of normal subjects to perceive the position of their body co-ordinates has been studied in many experiments. Particular emphasis has

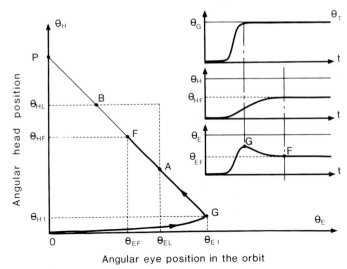

Angular eye position in the orbit

Fig. 4.10 A hypothesis accounting for the non-linearity of the eye and head position signals.

Diagrams on the right represent, from bottom to top, eye position in orbit (θ_E), head position (θ_H) and gaze position ($\theta_G = \theta_E + \theta_H$) as a function of time during a co-ordinated eye–head movement. The gaze is already at target at point G. Between points G and F (the end of the head movement), the eye is driven back in the orbit, due to vestibulo-ocular reflex. This compensatory eye displacement does not affect gaze fixation.

Diagram on the left represents the phase plane of head position vs eye position in orbit. Note that eye position changes rapidly up to point G, and that beyond this point the eye displacement in orbit is equal and opposite to the head movement (at point F). The hypothesis assumes that the eye position (θ_{EF}) and the head position θ_{HF} corresponding to point F represent the optimal configuration for a correct encoding of the corresponding eye and head position signals. If the head movement is too small (or does not occur because the head is blocked), eye position in orbit will be too excentric (e.g. beyond point A corresponding to the limiting value θ_{EL}), and poorly encoded. Similarly, if the head movement is too large (e.g. beyond point B) head position with respect to the trunk will reach a value θ_{HL} beyond which it is poorly encoded. (From Biguer *et al.* 1984.)

been put on perception of the vertical (see Howard 1982). The present section, however, is restricted to the perception of body midline or sagittal axis. Body midline is a functional axis which segments space into left and right sectors. Its perception is therefore directly relevant to the problem of spatial localization of objects in extrapersonal space. In fact, there are interactions between perception of the sagittal body axis and perception of the vertical. Subjects with different degrees of head and/or body tilt make errors in attempting to place a luminous line in conformity with the physical vertical: the error is in the direction of the head tilt, and increases with the amount of tilt (the so-called Aubert phenomenon). Mittelstaedt (1983) suggested that the subjective vertical was shifted toward the direction of the person's long axis as a result of competition between what he calls the 'idiotropic vector' and the gravity system, for dominance over a common reference. Because this idiotropic vector is not part of a specific sensory system (unlike the gravity system), it must be considered as a construct generated by an autonomous system of orientation.

One way to demonstrate the reality of the perception of the idiotropic vector is to ask subjects, in the absence of visual cues, to point their hand 'straight ahead', that is, in the direction where they feel that the middle of their body would project in front of them. By using this method, Werner *et al.* (1953) were able to show that subjects could locate their body midline with relatively good accuracy, though they consistently deviated from objective midline. When the subjects pointed without vision (they had their eyes closed), the mean pointing position was 3.8 cm to the right of objective midline. When the subjects fixated a small luminous target placed straight ahead (although they could not see their hand during point- ing) deviation of the mean pointing position was 1.4 cm to the right of objective midline. The larger error observed with the eyes closed might be explained by the fact that the subjects were standing upright during the experiment. In this situation, the absence of visual cues is known to increase the amplitude of body sway, which may reflect in inaccurate pointing.

The systematic pointing bias (to the right of objective midline in the Werner *et al.* experiment) is more interesting to consider. The fact that spatial orienting in normal subjects is biased towards one direction seems to have been known for a relatively long time. Schaeffer (1928) reported that blindfolded subjects walk in spiral paths when attempting to go in a straight line. Rightward spiralling was reported to be more frequent than leftward. Rightward bias was also reported in spontaneous head position in newborn infants (Gesell 1938; Liederman and Kinsbourne 1980). Other studies, however, seem to have clearly demonstrated the existence of a leftward bias in manual exploration of space. This phenomenon was first described by Bowers and Heilman (1980) in a task involving tactual explo- ration of a rod held in front of the subjects. A leftward bias was found for both hands in estimating the midpoint of the rod. The same was also true if the rod was placed in the subjects' right hemispace, but not in their left hemispace. A similar effect, referred to as 'left side underestimation' (LSU) was reported by Bradshaw *et al.* (1983), using either the tactual or the visual modes. Bradshaw *et al.* (1983) attributed LSU to the greater visuospatial processing power of the right hemisphere, which makes sub- jects see or feel the extent to the left of centre as larger than it actually is and consequently to objectively underestimate the left half of the rod (for a review, see Heilman *et al.* 1987). In agreement with these studies showing a leftward bias in line bisection tasks, Biguer and Jeannerod (unpublished

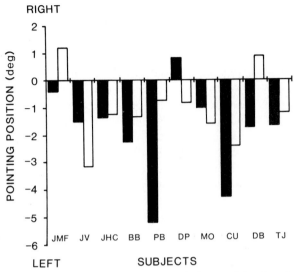

Fig. 4.11 Leftward bias in normal subjects pointing straight ahead. Ten subjects pointed with their right hand in the straight ahead direction, either in complete darkness (black bars) or with a visual target placed at the objective midline (white bars). They had their heads rigidly fixed. Note the consistent leftward bias. This effect (Biguer and Jeannerod, unpublished results) seems to be strongly dependent on experimental conditions, and particularly on the degree of head fixation and the side of the arm. In comparing this figure with Fig. 4.5, it can be seen that some subjects (JHC, CU) who pointed to the right in the experiment displayed in Fig. 4.5, now point to the left.

results) also found a consistent deviation to the left in pointing straight ahead. This study was conducted in ten right-handed adult subjects. They had their heads rigidly fixed by means of a bite plate. They used their right hands for pointing. Mean deviation from objective midline was 1.89° to the left when pointing was made in the dark, and 1.08° to the left when pointing was made toward a small light placed at objective midline (Fig. 4.11).

The experiments reported above, however, generally implied that the subjects could see the targets binocularly. The effects of monocular vision on the setting of straight ahead direction are more complex. According to Porac and Coren (1986), settings for straight ahead made with the right eye alone were biased to the right of objective sagittal median plane. Settings made with the left eye were biased to the left. Binocular settings fell between the directions of the two monocular settings. In addition, Porac and Coren noted that those of their subjects who were right sighters tended to set their apparent straight ahead systematically to the right (in either monocular or binocular conditions), while the left sighters deviated to the left. According to these authors, these results support the contention of a bias in egocentric localization toward the side of the sighting eye.

Demonstrating that the position of the body midline can be perceived as a reference for spatially oriented actions implies that it relies on signals independent from those generated by eye or head positions. Already in 1953, Werner *et al.* had reported a study where subjects attempted to point straight ahead while fixating a target located 15° right or left of objective midline. They found that pointing positions were shifted in the direction of the fixated target. No difference was found whether fixation was made by the eyes alone or with the eyes and the head aligned on the target. By contrast, when the subject's eyes or head were deviated to the right or to the left but the eyes were closed while pointing straight ahead, the mean pointing position was shifted in the direction opposite to that of eyes and/or head turn. In both conditions (fixation or eyes closed), the amplitude of the shift was much smaller than the actual eye or head turns.

The finding of opposite shifts of pointing positions whether a fixation target was available or not, is rather difficult to interpret. According to the authors, 'Since . . . when a fixated object is placed to the right the position of the apparent median plane shifts to the right, we assume that the innervation pattern of the organism correlative with that position is similar to that under left head torsion (eyes closed) . . . ' (p. 298), where the apparent median plane is also shifted to the right. To explain this similarity, Werner *et al.* (1953) postulated that in the case of fixation to the right, the head or eyes are maintained on target by 'active' forces, which are exerted in the same direction as the turn. In the case of eye or head deviation without fixation, however, 'counteractive' forces have to be exerted in the direction opposite to the turn, to counteract the normal tendency for the eyes and the head to remain aligned with each other. Accordingly, keeping the head deviated to the left with the eyes looking straight ahead in the dark, would in fact require a force to maintain the eyes deviated to the right with respect to the head. If the force generated to maintain the eye or head position is monitored and used to direct the pointing movements, these will be shifted to the right in both cases.

Further findings reported by Biguer and Jeannerod (unpublished results) only partly confirmed the Werner *et al.* interpretation. The same right-handed subjects as in the above-mentioned experiment, with their head fixed in the sagittal position, were requested to point straight ahead (with their right hand) while looking at targets located 5°, 10°, or 15° right or left of their objective midline. Systematic pointing errors were found, such that subjects located the straight ahead direction too far in the direction opposite to the gaze deviation. For example, eye fixation at a target located 5° to the right of objective midline yielded to a 4° mean pointing error to the left. An error of a similar amplitude to the right occurred in the case of eye fixation to the left. These errors did not change in sign but decreased significantly in amplitude when eye fixation was directed at more remote targets (e.g. 10° or 15°) (Fig. 4.12).

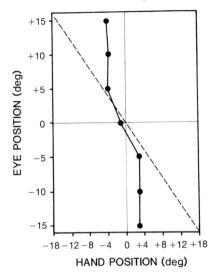

Fig. 4.12 Competition of the eye position signals and signals generated by the idiotropic vector for the determination of the straight ahead direction.

Head-fixed subjects pointed straight ahead while fixating visual targets located at midline or the right (+ signs) or to the left (− signs) of their body midline. When visual fixation was at midline, subjects pointed slightly to the left of midline (see Fig. 4.11). When the fixation target was at + or −5° subjects tended to exploit the eye position signals in computing the position of their idiotopic vector. This tendency disappeared during fixation on more remote targets. (From Biguer and Jeannerod, unpublished.)

A possible interpretation to these findings is that deviations of the line of sight from the head-trunk axis remained unnoticed because of the weakness of the eye position signals. Therefore, when subjects were to indicate the straight ahead direction, they logically erred in the direction opposite to the eye turn because, if the eye deviation was not perceived and the fixated target was perceived at midline, the head–trunk axis (the idiotropic vector) had to be perceived as deviating from its viridical position by the same amount as the amplitude of the eye deviation and in the opposite direction. This interpretation accounts only for part of the results, however. If, according to the above reasoning, subjects actually ignored signals generated by the orbital position of their eyes, their perceived straight ahead (as indicated by pointing) should have been negatively correlated to eye position. Figure 4.12 shows that this was not the case and that instead, pointing deviations were systematically smaller than eye deviations. This result indicates that both the eye position signals and signals generated by the idiotropic vector compete for the determination of the straight ahead direction. When the angle between the eyes and the head–trunk axis is small (e.g. ± 5°), eye position signals are too weak to be used as an index of the permanent eye deviation. When this angle increases, eye position signals become perceptible and can be used for detecting the deviation of the eyes with respect to the head–trunk axis, so that pointings get closer to the actual straight ahead. This hypothesis of a competition between two mechanisms for the determination of the straight ahead direction is quite distinct from that of Werner *et al.* (1953), which postulated the existence of counteractive forces to maintain aligned eye–head–trunk postures.

4.3.2. A note on the deviation of straight ahead after prism adaptation

A similar type of reasoning was used to account for effects on the direction of the subjective straight ahead resulting from prism adaptation, first reported by von Helmholtz in 1866. Typically, if an object is placed directly in front of a subject while he is wearing left-displacing prisms, he errs to the left in attempting to touch it. Adaptation will consist, under visual guidance, in redirecting arm movements in order to reach for the viridical position of the object. Accordingly, at the end of the adaptation period and after the prism is removed, the hand will err, now to the right of the actual position of the object. Von Helmholtz's interpretation of the phenomenon was that it is not the muscular feeling of the hand and arm which is at fault nor the judgement of their position, but the 'judgement of the direction of gaze'. Remember that he thought that knowledge of the direction of gaze axis did not depend upon the actual position of the eyeball or the tension of the extraocular muscles, but simply upon the 'effort of will involved in trying to alter the adjustment of the eyes'.

More recent prism experiments have confirmed von Helmholtz's hypothesis. (For a critical account of the straight ahead measurements in the context of this type of experiments, see Harris 1974). Kalil and Freedman (1966) have examined eye-in-head position in subjects following prism adaptation during a brief period (4 min). The subjects were placed in the dark with the head fixed in alignment with their trunk, and were requested to look straight ahead. Eye position measured photographically was found to be deviated in the direction of the previous optical displacement (see also Craske 1967; McLaughlin and Webster 1967). Similarly, Lackner (1973) had subjects adapt to a leftward visual displacement by wearing laterally displacing prisms. During the adaptation period, which lasted for 75 min, the subjects had to move actively and to interact with objects in their visual surround. Following adaptation, their resting eye position (measured with an after-image technique) was found to be deviated in the same direction as the optical displacement (i.e. to the left, as expected). In a subsequent experiment using the same subjects, Lackner (1973) measured the perceived head position following a 75 min prism exposure and producing leftward optical displacement. After adaptation the subjects, in the dark, were requested to place their head so that they felt it aligned with their trunk. In fact, they rotated their head too far to the left. Accordingly, when the subjects had their head objectively fixed in the median sagittal plane, they felt it to be turned too far to the right. Lackner logically concluded from these results that the head and eyes are behaving as a unit; during adaptation, the entire system is rotating with respect to the trunk in the direction of the optical displacement and a new reference position is established for the median plane, with the consequence that the subjects will feel their median plane in a position different from their objective straight ahead.

Although these results represent a positive test for von Helmholtz's hypothesis, they do not necessarily indicate that adaptation to visuomotor conflicts produced by prism exposure relies only on a change in perceived eye and/or head positions. Courjon *et al.* (1981) have explored this point by exposing normal subjects to a conflict between visual and visuomotor cues. Subjects saw monocularly a horizontal line, physically tilted by 20° or 40°, but appearing horizontal as a result of optical tilt in the opposite direction by rotating prisms. When they attempted to move their hand along the line, they experienced a slanted movement in conflict with the perceived visual orientation of the display. Exposure to this situation, however, rapidly produced adaptation, so that the hand appeared to move horizontally, now in conformity with the visual orientation of the line. The subjects were then immediately tested for their perception of the subjective vertical by stopping a line rotating around their gaze axis when they felt it was in conformity with the physical vertical. As a result of adaptation the vertical appeared to the subjects to be rotated in the same direction as the optical tilt produced by the prisms (see also Mikaelian and Held 1964). In the same experiment, the position of the eye in the orbit was examined by a photographic method. Eye position also appeared to be modified by exposure to the conflict, i.e. an ocular cyclotorsion was observed. Although this cyclotorsion was, in most subjects, in the same direction as the perceptual effect on the subjective vertical, no significant correlations were found between the two (Table 4.1). This result indicates that perceptual effects (such as a displacement of the subjective visual co-ordinates) and oculomotor changes produced by exposure to visuomotor conflicts, may co-exist without being causally related.

The results reviewed in this section are clearly in favour of neurophysiological signals documenting the idiotropic vector. It remains difficult, however, to determine the degree of independence of these signals with respect to those related to eye-in-head and head-on-body positions.

4.3.3. Asymmetrical behaviour following unilateral lesion of the nervous system

The notion of a body (or egocentric, or idiotropic) reference seems therefore clearly needed for explaining the directional coding of movements. This reference can be conceived as an equilibrium position between signals

Table 4.1. Effects of exposure to a visuomotor conflict produced by sliding the hand along a contour viewed through rotating prisms. Prism-induced tilt of the contour was obtained by setting the prisms at 20° or 40°. The effects were measured on subjective perception of the vertical and on ocular cyclotorsion (measured photographically).

Subject	Prism rotation					
	Visuomotor exposure				Visual exposure	
	20°		40°		40°	
	After-effect	Cyclotorsion	After-effect	Cyclotorsion	After-effect	Cyclotorsion
1	0.16	−1.30	2.94	0	1.55	0
2	1.80	2.50	1.84	−1.40	2.07	0
3	1.56	1.10	−0.28	0.87	−0.28	0
4	1.22	1.10	1.20	1.99	−0.07	0
5	2.74	1.00	1.65	1.69	0.88	0
6	2.01	2.20	2.08	2.89	0.32	1.12

The left side of the table (visuomotor exposure) displays the results for the two values of prism rotation. As expected the vertical was subjectively perceived as tilted to the right (after effect) in the six subjects, except one who had a reverse effect at 40° (− sign, after-effect to the left).

Ocular cyclotorsion was also observed, to the right in most cases. However, the amounts of tilt after-effect and of cyclotorsion were not correlated ($r = 0.35$, NS).

The right side of the table (visual exposure) displays the results of a control experiment, where subjects were exposed to the same amount of tilt while passively inspecting the contour. This exposure produced only a small after-effect and no cyclotorsion. (From Courjon *et al.* 1981.)

arising from both sides of space, and governing actions directed toward them. Pathological conditions where the equilibrium between the two sides is disrupted as a consequence of lesion of the peripheral or central nervous system, seems to be an interesting paradigm for studying this problem.

In invertebrates, the symmetry of orienting behaviour results from tonic inflow arising in bilaterally represented sensory organs. In these species, asymmetrical visual stimulation results in asymmetrical locomotory behaviour and postural abnormalities; unilateral lesion of these organs produces forced circling, such that the animal's orienting or locomotion is constantly biased toward the intact side. These effects have been attributed to asymmetrical repartition of 'visually induced tonus' (for a historical account of this theory, see Meyer and Bullock 1977).

In vertebrates, unilateral lesions of the central nervous system can produce similar effects. At the subcortical level unilateral lesion of the superior colliculus (Sprague and Meikle 1965; Flandrin and Jeannerod 1981), the thalamic intralaminar nuclei (Orem *et al.* 1973), the basal ganglia (Boussaoud and Joseph 1985) produce strong, albeit transient, behavioural asymmetries: animals tend to walk in circles toward the side of the lesion, and to ignore stimuli appearing on the other side. Spontaneous gaze posture is also deviated following such unilateral lesions. For example, cats with unilateral collicular lesion may show asymmetrical gaze posture at rest in a lighted environment (Flandrin and Jeannerod 1981), the gaze being directed toward the lesion side.

In the monkey, unilateral excision of associative frontal cortex produces strong ipsilesional forced circling (Kennard and Ectors 1938). Similarly, unilateral lesion of the posterior parietal cortex (particularly area 7) produces an ipsilesional bias of reaching movements (Faugier-Grimaud *et al.* 1978).

In humans, ipsilesional deviation of gaze posture is a common clinical finding during the acute stage following hemispheric lesion. Lesions restricted to the parietal lobe on one side often produce a systematic pointing or reaching bias toward the lesion side (Ratcliff and Davies-Jones 1972; Perenin and Vighetto 1983). Patients with unilateral neglect resulting from parietal lesion make large errors in pointing at an imaginary point in space in front of the middle of their chest. Instead of pointing roughly at the objective midline (as normal subjects would do), they err toward the non-neglected side, that is toward the lesion (Heilman *et al.* 1983). For a more complete account of the effects of parietal lesions on visuomotor behaviour, see Chapter 6.

According to Kinsbourne (1970), this abnormal behaviour suggests 'an imbalance between what are normally roughly equal and opposing orientational tendencies'. Kinsbourne proposed that each hemisphere subserved contralateral orientation by competing with the other for control of brainstem output mechanisms. This competition would be mediated by mutual inhibition across transverse commissures (Kinsbourne 1970). A well-known observation by Sprague (1966) supports this theory. Cats with unilateral lesion of the occipital lobe tend to neglect stimuli appearing contralaterally. This neglect is decreased by excision of the superior colliculus opposite to the cortical lesion, or simply by section of the intercollicular commissure. Sprague's interpretation of this phenomenon implied that the activity in the two superior colliculi became asymmetric after the cortical lesion, due to suppression of the excitatory cortico-tectal input on one side. As a consequence, the colliculus on the intact side exerted an exaggerated inhibitory influence on the other colliculus, and produced the neglect. Suppression of this inhibitory influence by section of the intercollicular commissure re-established some balance in activity between the two sides, and accounted for disappearance of the neglect. A similar reasoning might apply to neglect syndromes produced by cortical lesions in primates, and to the possible role of the corpus callosum in mediating cortico-subcortical inhibitory influences (see Watson *et al.* 1984).

One can only speculate on the nature of the orienting bias following unilateral lesions. One possible interpretation is that orienting bias occurs as one of several consequences of a higher order deficit, affecting central representation of the body reference system. If this interpretation were correct, one should be able to observe at lower levels of the nervous system, symptoms which would directly reflect compensatory reactions to the higher level dysfunction. The vestibular symptoms described following the same lesions as those that produce orienting bias might be explained along these lines.

It has recently been demonstrated that animals with such unilateral lesions present a spontaneous nystagmus in the dark, with the fast phase directed toward the lesion side. Vestibular stimulation in these animals induces asymmetrical responses, the gain being increased during rotation toward the lesion side and decreased during rotation the opposite way. This fact has been documented in the cat following unilateral lesion of the superior colliculus (Flandrin and Jeannerod 1981) and associative parietal cortex (Ventre 1985), and in the monkey following unilateral lesion of area 7 (Ventre and Faugier-Grimaud 1986) (Fig. 4.13). In humans, several authors have reported asymmetrical vestibular responses following unilateral frontal or parietal lobe lesions (Hécaen *et al.* 1951). As a rule, responses have been found to be stronger during rotation toward the lesion (e.g. Takemori *et al.* 1979).

These effects thus demonstrate the existence of a vestibulo-ocular bias in the direction opposite to the lesion side. In order to reconcile this phenomenon with the orienting bias following the same lesions, one might speculate that these unilateral lesions produce an illusory 'rotation' of the egocentric reference, somewhat as if the subject felt being constantly rotated toward the lesion side. In the absence of a visual reference frame (e.g. in the dark), this illusory 'rotation' would induce a compensatory vestibulo-ocular

Fig. 4.13   Effect of unilateral lesion of area 7 on vestibulo-ocular reflex in the monkey. Eye movements were recorded with an electromagnetic technique in one monkey (Chloe) before and after parietal lesion. Horizontal eye position was monitored at rest in the dark (a) or during sinusoidal rotation in the horizontal plane (b). The case represented is 0.03 Hz, $30°$, s^{-1}. From the recorded signals, VOR gain (c) and phase shift (d) were computed and plotted against frequency of rotatory stimulus, for each direction (CW, clockwise, CCW, counterclockwise). Before the lesion, note the symmetrical near-unity gain at all frequencies (open circles in c) and symmetrical phase-lead (open circles in d).

The lesion involved sub-pial aspiration of area 7 in the right hemisphere (see left side of the figure). Following this lesion, a spontaneous nystagmus with the fast phase to the right was observed at rest in the dark (a). VOR became asymmetrical with an increased gain during rotation to the right (CW) and a decreased gain during rotation to the left (CCW) (b, c, d). These effects were similar (albeit weaker) to those observed following a left hemilabyrinthectomy. (From Ventre and Faugier-Grimaud 1986, with permission.)

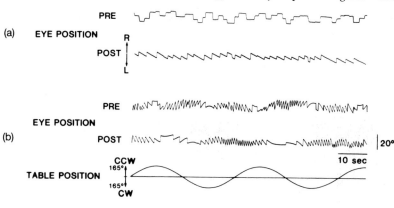

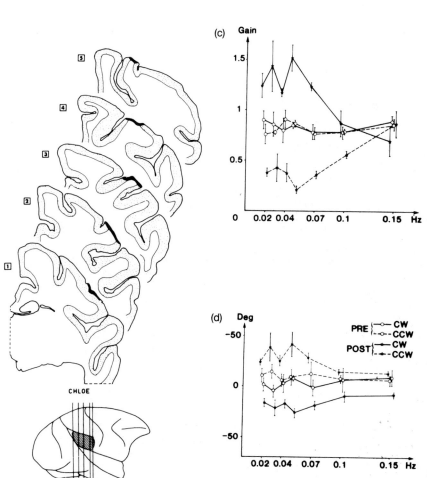

response toward the opposite side (Ventre *et al.* 1984). The notion of a compensatory vestibulo-ocular 'response' to explain the vestibulo-ocular asymmetry following unilateral lesions seems indirectly supported by the finding that vestibular stimulation in patients presenting unilateral neglect transiently reduces both their orienting bias and the extent of the neglected area (Rubens *et al.* 1985; Cappa *et al.* 1987; see also Silberpfennig 1941). In order to obtain the effect, it is necessary to stimulate the labyrinth on the same side as the cortical lesion, that is to reinforce the already existing vestibulo-ocular bias. This result therefore suggests that the vestibulo-ocular asymmetry observed following unilateral lesions is in fact a mechanism that opposes the orienting bias produced by the same lesions.

4.4. Conclusion: a hypothesis for spatial neglect

The above review has examined some of the neurophysiological mechanisms that operate at the junction between processing of incoming sensory signals and generation of corresponding motor outputs. This junction implies the delicate operation of transferring the position of object images on the sensory maps into a body-centred system of reference.

Neural signals for this transformation can be perturbed at several levels, with the common consequence of producing orienting bias, a systematic disorientation in one direction. In invertebrate and lower vertebrate species, where the sensorimotor transformation seems to be relatively direct, orienting bias may occur as a result of unilateral lesion or asymmetrical functioning at the receptor level. In higher vertebrates, this type of orienting bias can still be observed following unilateral labyrinthine lesion. Indeed, it remains to be demonstrated whether it would not occur also following unilateral sensory dysfunction in the visual or auditory modalities, for example.

Orienting bias produced by dysfunction or lesion at the central level is a more common finding in higher vertebrates. Direct alteration of the representation of body-centred space, or partial disconnection of this representation from its inputs (eye and/or head position signals, for instance) can be a cause of asymmetrical spatial behaviour. The region of space where position of objects cannot be encoded and represented in the proper system of co-ordinates, and toward which orienting movements can no longer be generated, is disregarded and neglected.

This interpretation therefore relates spatial neglect to the orienting bias. In essence, this is another version of theories like those of Kinsbourne (1987) and Heilman (see Heilman *et al.* 1987). It has the additional feature, however, of specifically predicting 'inattention' for part of the space in conditions where an orienting bias can be experimentally created.

5. The role of position sense in movement control

An essential aspect of motor control is the role played by signals related to the respective positions of limb segments prior to and during the movements. These signals, which generate 'position sense', arise from many sources, including not only peripheral (visual and somatosensory) receptors, but also central structures. Position sense thus cannot be seen as reflecting the activity of a 'simple' control device for execution and arrest of movements. In this chapter a more general conception will be developed, which will integrate position sense into the process of mapping an intended movement into an executed movement. Together with the internal representation of the goal and the motor program that initiates the neural commands to the muscles, position sense will be shown to be part of a complex, proactive rather than retroactive mechanism, which acts not only in the short term for steering movements, but also in the longer term in processes like motor learning and motor memory.

5.1. Position sense. A historical note

The concept of position sense progressively emerged in the middle of the last century, when it was realized that a considerable amount of sensory information could arise from muscles during limb movements, and signal these movements to the brain. The early contributions of Charles Bell and Charlton Bastian to what they called 'muscle sense' or 'muscular sense', has been reviewed by Jones (1972). Charles Bell's conception was particularly interesting in that it was an open conception. 'At one time,' he said, 'I entertained a doubt whether this [the consciousness of the positions of our limbs] proceeded from a knowledge of the conditions of the muscles or from the consciousness of the degree of effort which was directed to them in volition' (quoted by Phillips, 1986, p. 4). Those are the terms of a debate which is still continuing today.

The two opposite conceptions of position sense (whether it is thought to arise from afferent or from efferent signals) were deeply rooted in classical nineteenth century physiology. The afferent conception, before it was heralded by Sherrington and became the dominant theory (see below) was considered by many authors as unsatisfactory because of lack of experimental evidence. Waller (1891), for example, said that muscular sense, 'in the sense of centripetal process *from muscle* is not supported by any direct

171

proof, and so long as the alternative hypothesis of expanded energy in "motor" centres is not disproved, it is not possible to admit that the feeling is entirely of peripheral origin, nor that the muscular contribution is the predominant factor among its peripheral constituents' (p. 243).

The hypothesis of 'expanded energy' as a signal for position sense had strong proponents. As Bain stated: 'The sensibility accompanying muscular movement coincides with the *outgoing* stream of nervous energy, and does not, as in the case of pure sensation, result from any influence passing inwards by incarrying or sensitive nerves' (quoted by Lewes 1879, p. 23). Bain thought that there was a feeling of the exerted force and that this feeling was the 'concomitant of the outgoing current by which the muscles are stimulated to act'. This theory originated from several sources. One of them was clinical observation of patients with complete anaesthesia of one limb (including the loss of sensations generated by passive displacement), who were still able to produce voluntary movement with that limb. According to Duchenne de Boulogne (1855), these patients still presented 'muscular consciousness' ('conscience musculaire'), although they had lost muscular sense, that is, sensations generated by muscular activity. Duchenne was therefore led to the interesting conclusion, that muscular consciousness can exist independently of muscular sensations. This point was clearly formulated by Lewes (1879) who introspectively distinguished between the 'motor feeling' accompanying the active contraction of a muscle and the 'sensation' generated by this contraction. Lewes thus considered that the complex experience arising from a voluntary movement was the sum of both the 'Sense of Effort' and the 'Sense of Effect'. Along with Duchenne de Boulogne, Lewes in fact believed that both senses were of muscular origin: the sense of effort, he thought, was produced by impulses reaching the motor centres antidromically through the motor nerve during muscular contraction (whether this formulation anticipates proprioception is only a matter of speculation); the sense of effect was due to deformations of muscles, tendons, and skin during the movement.

A strong support for the efferent conception of position sense was the theory of the 'sensations of innervation' postulated by H. von Helmholtz and W. Wundt, both pupils of J. Müller. This theory held that sensations do not necessarily arise from peripheral sense organs, but that efferent activity leading to movement can be perceived as well. It therefore directly opposed the orthodox conception of W. James (1890) who considered afferent feelings as the only possible source of information for building an 'idea' of a movement. According to Wundt (1892; see Ross and Bischof 1981), the sensations of innervation arose from intracerebral intermediary fibres directly connecting the motor centre producing the movement to the corresponding sensory centre influenced by this movement. One of the key arguments for this theory was the description of illusions of movement experienced by patients with limb or eye muscle paralysis. During attempts to move a paralysed eye, for instance, these patients experienced instability of the visual world, as a result of central monitoring of the exaggerated 'effort of will' produced to displace the eye against the paralysis (see von Helmholtz 1866). Since there was no actual movement produced, the perceptual effect could not be due to changes at the muscular level. For a more complete account, see Chapter 4.

The debate on eye position sense, however, seems to have obscured the main issue of the contribution of sensations of innervation in position sense in general. The eye is a special case: it has little or no position sense, and the role of extra-ocular muscle proprioception for this purpose is controversial or even denied. This debate may have rigidified the terms of the afferent and efferent theories of position sense, which were often considered as mutually exclusive. In fact, at least for what concerned limb position sense, the complementarity of the two mechanisms was recognized early on. An example of a 'hybrid' conception was that advanced by Wundt (1892), who considered that the effects of the two mechanisms could only be separated out when ' . . . partial or total muscular paralysis had disturbed or entirely destroyed the other muscle-sensations which are peripherally excited' (quoted by Ross and Bischof 1981, p. 323), but that in normal conditions, the two co-exist.

In the following sections, the constituents of position sense will be reviewed in the light of these two opposite (though complementary) conceptions.

5.2. Afferent contributions to position sense

The physiological mechanism of the afferent component of muscular sense was established by Sherrington, who discovered the intramuscular sensory endings (Sherrington 1894), demonstrated their role in both initiation and control of movements (Mott and Sherrington 1895), and later coined the term proprioception to replace that of muscular sense. The role of proprioception in contributing directly to position sense was firmly held by Sherrington, who considered that the perceptions of posture, passive movement, and resistance to movement were mediated by proprioceptive mechanisms (see Sherrington 1900).

5.2.1. Deafferentation experiments

As already examined in Chapter 1, theories on the control of movements have been greatly influenced by the results of deafferentation experiments in animals, which were first reported about 90 years ago by Mott and Sherrington (1895). These authors had found in monkeys that a somesthetically deafferented limb (by section of the dorsal spinal roots) became virtually useless, and could only produce awkward movements when the animal was forced to use it. Hence their conclusion was that movements owed much to the periphery for what concerned both their initiation and their execution.

The classical Mott and Sherrington findings were later replicated by others. Lassek (1953) observed in 12 monkeys that dorsal root section corresponding to one upper limb prevented the use of that limb for at least 4 months following the operation. Lassek and Moyer (1953) had observed a slight recovery in one animal operated at the age of 4 months. Immediately after surgery the deafferented arm remained unused. About 6 months later, the animal became able to stretch its arm in the direction of food but finger movements were apparently lost. Twitchell (1954) reported closely similar results, with the additional information that sparing one of the dorsal roots innervating the arm was sufficient for maintaining a quasi-normal motor function.

Another line of experimenters, however, also using monkeys, reported results that were in striking opposition to those obtained by Mott and Sherrington and their followers. Munk (1909) clearly observed recovery of motor function, including for grasping small pieces of food, in monkeys with deafferented limbs. More recently, Knapp *et al.* (1963), Taub and Berman 1968, Bossom and Ommaya (1968), and Liu and Chambers (1971) all reported that monkeys with both arms deafferented were able to perform adequate reaching and grasping movements within 2 weeks of operation. Movements were first ataxic and awkward, but tended to improve with time, provided visual feedback from the limb was available. According to Taub (1976), virtually normal prehension may develop in these animals: first, the hand is used to sweep objects back along the floor; later a crude grasp is made by all four fingers in unison without participation of the thumb; finally, a primitive pincer grasp develops, usually a few months after surgery.

These differences in the results of deafferentation experiments have been explained in several ways. Bossom and Ommaya (1968) have stressed the fact that motor pathways can be easily damaged during rhizotomy: when this occurs, as they observed in one of their animals, no recovery seems to be possible. It is also known that animals with bilateral deafferentation recover more rapidly than animals with deafferentation limited to one arm. Taub (1976) suggested that unilaterally deafferented animals may learn not to use the deafferented arm or even develop (intraspinal ?) inhibition of that arm. This hypothesis seems to account for the observation made by Knapp *et al.* (1963) and Bossom (1972, 1974), that unilaterally deafferented animals may recover movements with their deafferented arm to the same extent as bilaterally deafferented animals, provided their normal limb remains permanently attached.

Although the above-mentioned deafferentation experiments provided general information about persistence of motor abilities in the absence of proprioceptive feedback (and in the absence of any feedback, when vision of the limb is prevented during the movement), the problem of which movement parameters are preserved following deafferentation, and to what extent these residual abilities can achieve real accuracy, required more refined testing. This has been attempted by Bizzi and his colleagues in monkeys with bilaterally deafferented arms. Before operation, animals were trained to point one arm at luminous targets; no vision of the limb was allowed. At times during the experiment, the initial position of the forearm was displaced just prior to onset of muscular discharges initiating the movement. In spite of this procedure, normal trained animals always reached the target with precision. After deafferentation, not only could the monkeys initiate the movement, but they could also reach targets with relative accuracy. Displacement of the forearm prior to the movement did not alter significantly pointing accuracy of these trials compared to non-perturbed trials (Polit and Bizzi 1979). Bizzi concluded from these findings that simple, single-joint movements depend on neural patterns that are programmed prior to movement onset, and that no feedback is required for achieving the final limb position determined by the program. Bizzi's observation is not unique. Relatively accurate pointing at visual targets has been demonstrated in deafferented monkeys by Bossom and Ommaya (1968), Liu and Chambers (1971), and Taub *et al.* (1975). The latter authors have shown in four deafferented animals that during pointing at visual targets (with or without visual feedback from the moving limb), constant error remained within the range of values observed in normal animals tested in comparable conditions of visual feedback. Variable error, however, was largely increased in the deafferented animals, particularly in the condition where visual feedback was not available.

Finally, recent data obtained from human subjects following pathological deafferentation at the peripheral level have confirmed the experimental findings, and in addition have set the limits of the role of sensory feedback in the control of movements. Rothwell *et al.* (1982) have studied one patient (G.O.) suffering a severe peripheral sensory neuropathy, which had resulted in loss of sensations from the four extremities. This patient was subjected to extensive testing of his motor skills. Clinically, light tough sense, vibration sense, and temperature sensation were impaired or totally absent at the level of both hands. In spite of this deficit, the subject was able to perform a wide range of motor tasks even with his eyes closed. Tapping, fast alternating flexion extension movements, and drawing figures in the air were easily executed by the subject using only his wrist or fingers. His performance rapidly degraded, however, when he was asked to repeat several times the same simple movement with the eyes closed. Electromyographic (EMG) recordings of thumb flexor and extensor muscles were performed in this patient in order, in the terms of Rothwell *et al.*, to 'assess the accuracy with which a central motor program could be selected and performed in the absence of sensory information' (p. 527). EMG discharges

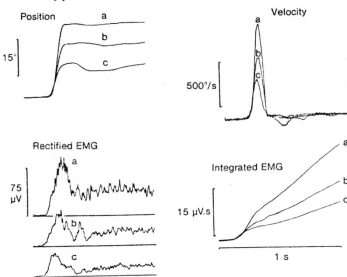

Fig. 5.1 Rapid thumb flexion in the absence of visual feedback, in a patient with complete anaesthesia of the hand, due to peripheral sensory neuropathy. The subject was able to move his thumb fast and accurately to one of three different end positions (a, b, c upper left). His performance in the absence of vision was similar to that in control trials with vision, and was little different from that of normal subjects. Note that peak velocity was scaled to movement amplitude, as it is in normal subjects. Also note a clear invariance of movement duration despite increase in amplitude and velocity (upper right). These properties are reflected by graded increase in EMG activity of the agonist muscle (flexor pollicis longus, lower row) during the same three movements (a, b, c). This result is indicative of integrity of the motor programs for simple single-joint movements in this deafferented patient. See Chapter 1. (From Rothwell *et al.* 1982, with permission.)

during flexion–extension of the thumb were found to be closely similar to those observed in normal subjects. The typical triple burst pattern of agonist and antagonist muscles was clearly present. The subject was also able to learn, with vision, different thumb positions, and to accurately reproduce them by flexing his thumb without vision. In that case, duration of the movement remained constant for the tested amplitudes of thumb flexion. Increase in movement amplitude was obtained by increasing the amplitude of the flexor EMG burst within an approximately constant duration (Fig. 5.1), as would be observed in normal subjects. Preservation of time invariance in executing movements without proprioceptive feedback is thus a strong argument in favour of pre-organization of motor output. The same patient was also able, without vision, to select (but not to maintain) a prescribed level of force by exerting an isometric thumb flexion against a lever.

These findings were confirmed and expanded by Sanes *et al.* (1985) in a

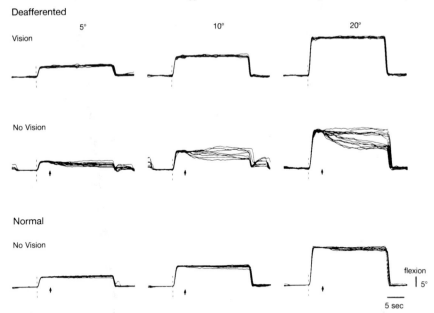

Fig. 5.2 Maintenance of a steady posture of the wrist against an elastic load, in a patient with a complete anaesthesia of the arm due to a periphery sensory neuropathy. The patient was requested to rotate his wrist rapidly in order to reach target positions at 5, 10, and 20°. In the presence of visual feedback (upper row) movements were executed rapidly, the target position was reached, and was maintained against the load. When visual feedback was turned off (arrow, middle row) the position could not be maintained and the wrist was pulled back in the direction of the load. For comparison, see the performance of a normal subject in the same condition (lower row). For each target, 8–13 trials have been superimposed. (From Sanes *et al.* 1985, with permission.)

group of patients also suffering peripheral sensory neuropathy. The patients could execute repetitive flexion–extension movements of the wrist with the normal alternating EMG pattern except that, beyond a certain frequency the intervals between EMG bursts tended to disappear. As a rule, EMG during movements in these patients revealed an important degree of co-contraction of agonist and antagonist muscles. The patients were also able to maintain a steady posture with their deafferented limb segments as long as visual control of the limb was possible. Similarly, they were able to correctly make a movement against an elastic load and to maintain the corresponding posture. Without visual control, however, relatively large errors were made, position could not be maintained, and the limb drifted back to its initial position (Fig. 5.2).

In conclusion of this review of the effects of somatosensory deafferentation on motor control, it appears that kinaesthetic feedback is not likely to be necessary for movement initiation and execution. This is not to say that movements can be entirely controlled by preset commands. Deafferented animals are unable to reach accurately for targets if they have to make movements involving several joints, or if the resting position of their limb is changed prior to the movement (see Chapter 1). In humans, good performance with deafferented limbs, especially at the level of distal segments, has been demonstrated only with simple, single-joint (usually flexion–extension) movements. This point was clearly stated by Rothwell *et al.* (1982). Their patient G.O. was dramatically impaired for natural movements performed in everyday life, due to his inability to maintain a constant motor output, to produce motor sequences, and to perform co-ordinated movements involving several joints. These results point to the contribution of limb position sense, not only for controlling the execution of movements, but also for providing proactive information at the programming level.

5.2.2. Time for proprioceptive feedback to be effective

The second critical issue for discussing the role of somatosensory feedback in motor control is the time needed for proprioceptive feedback to be effective in controlling an ongoing movement. This delay, referred to as the kinaesthetic reaction time, was measured as early as 1948 by Vince. In Vince's experiment a subject was trained to make accurate movements in pulling a handle against a spring, at an acoustic signal. In some trials the stiffness of the spring was unexpectedly increased. The additional time needed by the subject to react to this new situation and to move the handle was estimated at 160 ms (Vince, 1948). Another classical estimate of kinaesthetic reaction time was made by Chernikoff and Taylor (1952). In this experiment, the relaxed arm of a subject was suddenly dropped by releasing an electromagnet. The subject was instructed to stop his falling arm as soon as possible. The change in acceleration of the arm was taken as the indicator of the subject's response. A mean value of 118.9 ms was found (Chernikoff and Taylor 1952). These early results were later confirmed by Higgins and Angel (1970) in an experiment where subjects had to hold stationary a joystick against unexpected pulls from a load. The load produced a passive movement of the arm that the subject had to stop: the onset of joystick deceleration was taken as the index of subject's response to the perturbation. A mean value of 107–169 ms was found according to subjects. Other authors also have obtained values within the same range (e.g. 90–100 ms, Hammond 1960; Lee and Tatton 1975) or even shorter (in the range of 70 ms in some subjects, Craggo *et al.* 1976; Evarts and Vaughn 1978). These results indicate that movements lasting less than some 100 ms, or the first 100 ms or so of any movement, are likely to be under the exclusive dependence of central factors, and owe nothing to peripheral feedback (see a similar discussion about visual feedback in Chapter 3).

The problem of the nature of the mechanism compensating for the perturbation mechanism (is it of 'reflex' or 'voluntary' origin?) is also important to consider. There have been discussions on whether these motor responses corresponded or not to the M3 component of the EMG response to stretch (Lee and Tatton, 1975). This component has been assigned a 'transcortical' origin, which would be more in favour of its participation in a voluntary than in a reflex mechanism (see Evarts and Vaughn 1978; Marsden *et al.* 1978).

This point was specifically addressed by Marsden *et al.* (1978) in an experiment where they measured the EMG response of the thumb flexor to a brief stretch. The subjects were instructed to hold their thumb flexed 10° against a constant load. The load could be suddenly increased, so that the thumb was pulled back to a new position in a time that could be varied from 13 to 45 ms. The stretch produced an early contraction of the muscle in about 25 ms, followed by secondary EMG bursts at 40 and 56 ms. These components were likely to correspond to spinal and supraspinal functional stretch reflexes. Finally, a larger and long-lasting burst was seen after about 118 ms following rapid stretches, and about 137 ms following slower stretches. The question arose of whether this late response to stretch actually reflected voluntary activity from the subject, and not just another reflex component. Marsden *et al.* (1978) have solved this problem by giving the subjects different instructions, that is, either to let the thumb go when the stretch came, or to resist as hard as possible. The effect of subjects' intent was reflected by the fact that the late component was significantly larger when the subjects were to resist than when they were told to let go. By contrast, instructions given to the subjects did not influence the size of the earlier components (Fig. 5.3).

The adaptive nature of mechanisms used to compensate for the effects of perturbations was further demonstrated by Newell and Houk (1983). Reaction times of motor responses to loading or unloading the arm at the wrist level were measured by taking the change in force as an index for the responses. When the perturbation was consistently applied in the same direction (e.g. loading) the mean reaction times ranged from 135 to 181 ms, depending on the subject. In the situation where the direction of the perturbation was unknown to the subjects, mean reaction times were significantly longer (158–234 ms). This finding indicates that such compensatory responses are not stereotyped and involve decision-making processes to the same extent as responses to stimuli in other sensory modalities.

5.2.3. The respective role of proprioceptive receptors in position sense

Deafferentation and reaction time experiments reported in the above two sections carried only indirect information on the contribution of proprio-

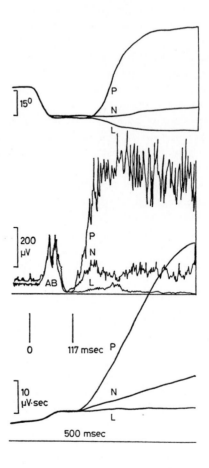

Fig. 5.3   The effect of the subject's response to stretch of a contracting muscle. The subject held the thumb flexed to 10° against a force of 2 N. At 50 ms after the sweep commenced, the thumb was stretched. In separate trials, the subject was instructed to either merely hold the thumb in a steady position (N), or to relax the thumb (L), or to flex as hard as possible (P) the thumb perceiving the stretch. Each record is the average of 48 trials. The upper records show superimposed the angular position of the thumb, the centre records show the full-wave rectified EMG responses recorded via surface electrodes from flexor pollicis longus, the lower records show the full wave rectified and integrated EMG record from flexor pollicis longus. Note consistent change in EMG activation around 177 ms following application of the stretch. The amplitude of the EMG change differed according to the instruction given to the subject. By contrast the earlier components of the response to stretch (AB) were not affected by the instructions. (From Marsden *et al.* 1978, with permission.)

ception to position sense. Direct demonstration of this contribution requires experimental situations where position sense can be tested during specific stimulation or exclusion of each type of proprioceptive receptor. As a rule, these experiments should also exclude active movements of subjects' limbs in order to isolate the afferent mechanisms from possible contamination by efferent processes.

(a) Positive arguments on the role of joint receptors

Until about 15 years ago, mechanoreceptors located in and around the joints were considered as the only afferent source of position sense. This belief was supported by neurophysiological studies on anaesthetized animals, which had clearly established position coding by joint receptors. These receptors were found to adapt very slowly and to keep a constant firing rate when a steady joint position was maintained (Boyd and Roberts 1953). They were also found to project at the cortical level, a characteristic compatible with their involvement in conscious experience, for giving rise to subjective position sense (see Rose and Mountcastle 1959).

Psychophysical experiments also contributed positive arguments as to the role of joint receptors. Provins (1958) measured in normal subjects the passive movement detection threshold at the metacarpophalangeal joint of the index finger. During the control session subjects were able to detect joint rotations of about 6°. When the joint (and the overlying skin and tissues) was anaesthetized by local instillation of xylocaine and the subjects kept their finger relaxed, the detection threshold was raised to about 15°. In order to test the possible role of muscular receptors in detecting passive displacement, the same experiment of passively rotating the joint was carried out with the anaesthetized index finger isometrically flexed or extended. The experimental evidence suggested that in this condition, tensing the muscle did not lower the displacement threshold 'because', Provins assumed, 'no adequate sensory information was derived from receptors in the muscles tensed' (1958, p. 63). Hence the logical conclusion that there is 'no justification for suggesting that appreciation of active limb movement or joint position with the muscles tensed is based on information derived from a source different from that used in the appreciation of passive movement'. (p. 64).

Discharge of slowly adapting joint mechanoreceptors was thus considered by theorists like Adams (1977) as the main source of position feedback during movements and postural attitudes. Based on neurophysiological findings in the cat (Boyd and Roberts 1953), the role of these receptors was considered by Adams as twofold. First, they regulated dynamic timing of the limb, because their discharge rate during a movement (dynamic frequency) was apparently related to limb velocity. Second, they also regulated static timing, or the sequence of the action, because, after a movement was completed, their maximum dynamic frequency decayed until it reached a steady-rate level (static frequency), characteristic for the new position of the limb. The rate of decrease in activity of the dynamic burst thus represented, according to Adams, information for triggering the next movement in the sequence.

(b) Negative arguments on the role of joint receptors

Other experimental data, however, came as a disappointment for those who saw joint receptors as the origin of position sense. Further careful neurophysiological experiments, also in anaesthetized animals, showed that most joint receptors, although they signalled the extreme range of joint rotation, stopped discharging when the joint was in its mid position (for references see Matthews 1982). This finding is hardly compatible with a role of joint receptors in position sense because, as stressed by Matthews, 'If there is no signal, then there can be no information about what is happening to the joint; yet, kinaesthesia demonstrably does not vanish when a joint is in the middle of its working range' (1982, p. 193).

Another negative argument comes from behavioural observations, which tend to show that position sense can survive elimination of joint input. For instance, Clark *et al.* (1979) demonstrated that intracapsular anaesthesia of the knee does not impair detection of absolute positions passively imposed at that joint, even if one takes care to move the limb so slowly that no movement is perceived. Also, replacement of joints by prosthetic material in patients does not abolish position sense. Following such operations the patients are able to perform position detection tasks within normal limits (Cross and McCloskey 1973; Grigg *et al.* 1973; Kelso *et al.* 1980).

Finally, other behavioural findings are in direct contradiction to one of the main properties of joint receptors, their virtual lack of adaptation. Brouchon and Paillard (1966) and Paillard and Brouchon (1968) requested blindfolded subjects to point with one hand at the position of their other hand, after it had been passively displaced by an experimenter. They found that the subjects made relatively large errors in estimating the position of their passively displaced hand, and that these errors increased as a function of time when pointing was delayed after the displacement. This result indicates that the steady-state discharge provided by the joint receptors cannot be used as a signal for position sense. It remains that position of the joints might be read from the dynamic frequency of the same receptors, rather than from their static frequency. The degradation of the estimation of hand position following a displacement might follow the rapid degradation of the dynamic discharge of the joint receptors (Boyd and Roberts 1953). However, the Paillard and Brouchon findings could not be replicated by other authors. Lee and Kelso (1979) found that errors in estimating passive limb positions were similar for durations of target position coinciding with the dynamic (3 s, 6 s) or the static (15 s, 30 s, 150 s) phases of the joint receptor discharge following a movement. In addition, they found that the errors increased with increasing joint angles. (For another failure to replicate Paillard and Brouchon's results, see Jones 1974). Lee and Kelso (1979) concluded from their findings that limb position sense cannot be predicted from the known physiological properties of the joint receptors.

(c) The role of muscle spindles

The Sherrington tradition held that the muscle spindles were unsuited for signalling position of joints. The main reasons for this belief were twofold. First, because muscle spindles can modify their own length due to their contractile structure, they were assumed to measure only the relative length of the muscles to which they are attached, although position sense would require measurement of absolute length. Second, muscle spindles were assumed to be 'insensient' and, therefore, to be unable to subserve conscious position sense (see Merton 1964). These assumptions were seriously questioned by many subsequent experimenters who, in fact, have now established an opposite conception of the role of muscle spindles in position sense.

Vibration experiments. The key experiment for this demonstration was the use of tendon vibration, a stimulus known to be specific for activating primary endings of the muscle spindles (Brown *et al.* 1967), in the context of human psychophysics. Thus, Goodwin *et al.* (1971, 1972) found that vibrating the biceps tendon at the elbow produced the 'consistent illusion that, even though it was still, the elbow was moving in the direction that it would have been if the vibrated muscle was being stretched; thus, on vibrating the biceps tendon, the arm was felt to be extending, while triceps vibration caused an illusion of flexion.' The explanation of this phenomenon, according to Matthews (1982), one of the authors of the initial finding, is that 'the excess Ia discharge was treated as if it were due to a gross muscle stretch, rather than to the minuscule stretch of vibration which has no counterpart in nature, and led to kinaesthetic sensations. These were referred to the joint, instead of the muscle itself, and used in the elaboration of the body image.' (1982, p. 199).

Eklund (1972) made a systematic comparison of limb position detection in normal subjects in conditions involving voluntary movement, passive displacement, or tonic muscular contraction elicited by vibration of the quadriceps (the so-called 'tonic vibration reflex', TVR). The subjects' task was to match the position of a moved, displaced, or vibrated leg, with that of the other leg. Matching leg position following voluntary movement was excellent. A clear undershooting was observed following passive displacement. By contrast, subjects overshot the leg position resulting from the tendon vibration reflex. Eklund's interpretation of these findings was that 'high receptor activity (as during TVR) gives an illusion of the quadriceps muscle being longer (the knee more flexed)' (1972, p. 610), whereas low receptor activity as during passive positioning with a relaxed quadriceps, gives an opposite illusion.

Vibration experiments have now produced a large body of data confirming the earlier findings of Goodwin *et al.* (1972). In addition, some authors seem to have been able to demonstrate that the apparent motion experienced by the subjects during the vibration is in fact treated by the central nervous system as if it were a physical movement. Lackner (1985) recorded eye movements in subjects attempting to fixate (in the dark) their fingertip while their biceps was vibrated. Many of the subjects exhibited pursuit-like eye movements in the direction of the apparent forearm motion. As both the subjects' head and vibrated arm were immobilized, this finding suggests that the changing apparent position of the arm was treated centrally as a physical change in forearm position. If a small light was attached to the finger, and the subjects were asked to fixate it during vibration, the light was seen to move in keeping with the perceived motion of the forearm. Of course, as subjects maintained steady fixation on the light, itself attached to the fixed arm, no ocular pursuit was recorded. According to Lackner (1985) the illusory forearm motion was greater in the first situation (where the eyes moved) than in the second situation (where they did not). He concluded that information about forearm position generated by muscle vibration and by changes in gaze axis interacted with each other for producing position sense.

The effects of vibrating neck muscles, fully described in Chapter 4 and only briefly covered here (see Biguer *et al.* 1986), can also be interpreted along the same lines. 'Head-fixed' subjects placed in a dark room were submitted to vibration of their posterior neck muscles on the left side. Their gaze was maintained immobile by a small fixation light placed in front of them. On vibration, subjects consistently experienced displacement of the fixation light, usually to the right, sometimes in other directions, depending on which muscle was predominantly stimulated. This result is thus even more striking than that of Lackner (1985), because subjects never reported illusory motion of their head or body. Instead, the proprioceptive signal produced by the vibration was directly translated into a perceived visual displacement and centrally treated as such. Indeed, subjects pointed by hand at the fixation light as if it were effectively displaced from its viridical position (Biguer *et al.* 1986).

Tendon-pulling experiments.   Demonstration of muscle spindle contribution to position sense was also provided by experiments in which surgically isolated tendons were stretched. At variance with the earlier work of Gelfan and Carter (1967), who had reported lack of subjective experience relevant to position sense during such manipulations, several authors have now reported feelings of movement on muscle pull. McCloskey *et al.*

(1983a) described an experiment made with the first author as a volunteer subject. The tendon of his extensor hallucis longus was surgically isolated and sectioned, and its proximal end fitted to a rod connected to an electromagnetic puller. The experimenters first noted that the subject could readily detect movements imposed on the interphalangeal joint of the toe after its extensor was cut. Then the toe was immobilized. As the tendon was pulled a sensation of plantar flexion of the toe at the interphalangeal joint was consistently reported. As the tendon was suddenly released, there was a sensation of dorsiflexion. With the tendon held fully extended, the toe felt plantar flexed. Vibration applied directly to the tendon also elicited a sensation of plantar flexion. It is remarkable that the sensations reported in this experiment were those of *joint rotation* and were not referred to the muscle itself. In other words, the sensory effects of stimulating muscle spindles were directly translated into position sense.

A similar experiment, also involving the author as a subject, was reported by Moberg (1983). The exposed tendon was that of the index flexor superficialis, which was left intact and pulled in the centrifugal direction. The author described sensations of extension of the index finger, but only when large pulls (encompassing the physiological range of movements for this tendon) were applied. Moberg concluded that his own sensations of movements could have been provoked by deformation of the skin in the region of the forearm and erroneously attributed to muscle spindle stimulation.

In fact, impressions of joint rotation during tendon pulling were confirmed in groups of patients tested during the course of hand surgery (Moberg 1983; McCloskey *et al.* 1983a). Different experimental conditions might account for the differences in interpretation between the two authors about the origin of their own feelings. As Phillips (1986) puts it 'it might be thought possible that the different functions of the hands and feet would be reflected in different degrees of responsiveness to the isolated spindle input' (p. 55). The toes have postural and locomotor functions, and the perception that can be aroused from their movements could depend more on the spindles than on the skin. The converse could be true for the fingers.

Both vibration and pulling experiments support the notion of an involvement of muscle spindles in position sense. In addition, the above findings are consistent with the results of neurophysiological studies demonstrating cortical projection of muscle afferents (Phillips *et al.* 1971; see Phillips 1986), a property that is admittedly required for mediating 'conscious' position sense.

(d) The mixed contribution of afferent mechanisms

The contribution of afferent mechanisms to position sense is thus a mixed contribution. The role of muscle spindles does not contradict that of joint, and possibly skin, receptors. In order to sort out the respective contributions of joint sense and muscle sense to position sense, Gandevia and McCloskey (1976) took advantage of an anatomical peculiarity of the hand. If the index, ring, and little fingers are held extended and the middle finger alone is flexed at the proximal interphalangeal joint, the terminal phalanx of that finger cannot be moved voluntarily. In this position, the joint is freed from effective muscular attachment. Position sense in response to passive displacement of the distal phalanx in this condition should reflect only joint sense, although the same passive displacements in a condition where the muscles are engaged should stimulate also muscle sense. Gandevia and McCloskey showed that combined stimulation of muscle and joint receptors greatly improved position sense, with respect to the situation where only the joint receptors were involved. This result is consistent with McCloskey's self-report on sensations of movement evoked by tendon pulling (i.e. in the absence of joint rotation). He reported his sensations as 'unusual', 'as if the toe is blowing back and forth in a breeze' (McCloskey *et al.* 1983a, p. 28), because they were unaccompanied by sensations of cutaneous distorsion and pressure in the toe itself.

Skin afferents indeed represent an important mechanism for position sense. Skin mechanoreceptors in the glabrous area of the hand have been shown to be massively influenced by isotonic movements of the fingers (Hulliger *et al.* 1979). It seems practically impossible to test whether selective exclusion of this mechanism (by anaesthesia of the skin alone) would alter or even abolish position sense, as Moberg (1983) suggested. This possibility seems to be ruled out by the Gandevia's observation following transcutaneous electrical stimulation of the low-threshold muscle afferents from the hand. This author was able to show that such a stimulation produced illusory movements of the fingers (Gandevia, 1985). As these illusions were not accompanied by changes of EMG activity of the corresponding muscles, nor by cutaneous sensations, they were likely to correspond to selective activation of fibres innervating muscle receptors.

5.3. Efferent contribution to position sense

5.3.1. Position detection following active versus passive movement

Comparison of subjects' accuracy in detecting their limb position following active or passive limb displacement, at first sight seems to represent a good way for sorting out the role of central commands in position sense.

It has been known for some time that active position of a limb yields to better detection of the position of that limb than does passive positioning (Goldscheider 1898; Lloyd and Caldwell, 1965). Paillard and Brouchon (1968) and Eklund (1972) showed that subjects pointing with one limb at the position of their other, passively displaced, limb markedly undershot. Undershooting disappeared when they had achieved the target position themselves. Similarly, Jones (1974) reported a study where blindfolded subjects either produced voluntarily forearm movements of various amplitudes (10°–60°) by moving one finger along a slide, or the same forearm displacements were imposed passively. Immediately after the hand had returned to its initial position, subjects attempted to voluntarily duplicate the movement. When duplication followed an active movement, almost no error was observed. When it followed a passive displacement, errors by undershooting were observed, and the scatter of duplicated amplitudes grossly increased.

Active and passive conditions were also compared for accuracy of limb position detection in another series of experiments using the eye–hand tracking paradigm. Steinbach and Held (1968) had shown that eye tracking of a visual target attached to the subject's arm was smoother and more accurately time-locked to the target when the arm was moved actively, than when it was passively displaced. Steinbach and Held thus suggested that smooth tracking in the active condition implicated feedforward information about the dynamics of the hand movement. Their finding, however, could not be replicated. Both Gauthier and Hofferer (1976) and Mather and Lackner (1980) found no difference in eye movements during the tracking of one's own limb in the dark, whether it was actively moved or passively displaced. Mather and Lackner concluded that, 'since active and passive movements of the hand were tracked with nearly identical efficiency', information about the command to move the arm played 'a minor role in specification of hand position for eye movement control' (1980, p. 313).

Even though active movements are superior for detection of static limb positions or for replication of previous positions, this does not necessarily demonstrate the contribution of efferent mechanisms in position sense. Active movements provide larger proprioceptive signals than passive movements. Both isotonic muscle contraction and the alpha–gamma linkage during voluntary movement result in an increase in the range of motion in which the spindles can respond, and in the overall volume of the response.

5.3.2. The perception of motor commands

In order to attribute a significant role to motor commands in the limb position sense, it is necessary to demonstrate that these commands can, in one way or another, be perceived by the subject. In the same way as in the case of eye position sense (see Chapter 4), experimenters have used subjective reports for assessing the contribution of this mechanism. In fact, the notion of perception of motor commands should be conceived in a broad sense, and not only referred to the production of subjectively detectable (and reportable) movements or changes in position. As stressed by McCloskey *et al.* (1983b), 'various terms, including sensations of innervation, sense of effort and felt will, have been used to refer to sensations said to arise directly from the internal action of motor commands' (p. 151).

Subjects presenting paralysis, following a stroke for instance, report that when they attempt to move their paralysed limb they may feel the exaggerated effort that they have to make. Mach (1906), in describing his own sensations following right hemiplegia, reported that when the paralysis was total, his attempts to move, although clearly perceptible, were not accompanied by a sensation of effort. Later, after recuperation had begun and the paralysis was less dense, each attempt was accompanied by the sensation of having the right hand and foot held down by heavy weights. The sensation of effort was thus perceived as a sensation of resistance which opposed movement. Similarly, Brodal (1973) reported that when he wished to move one of his paralysed extremities, it was ' . . . as if the muscle was unwilling to contract, and as if there was a resistance which could be overcome by a strong voluntary innervation. This force of innervation, is obviously some kind of *mental* energy . . . ' (p. 685).

More recently, Gandevia (1982) asked several paralysed patients to describe their sensations. Descriptions from patients with paralysis of central origin (e.g. hemiplegia) conformed to that of Mach (1906). They reported that attempts to move completely paralysed limbs produced no feeling of heaviness. When ability to move began to return, attempts were accompanied by a feeling of intense heaviness. Finally, the sensations of heaviness decreased as movements became easier and regained strength. By contrast, in patients with paralysis of peripheral origin, attempts to move the paralysed limbs were always associated with sensations of heaviness.*

Experiments involving transient paralysis in volunteer subjects have led to more ambiguous results. Several authors using ischaemic block or local injection of curarizing agents to produce paralysis of one hand, failed to observe sensations of movement during their attempts to move (Laszlo 1966; Goodwin *et al.* 1972, McCloskey and Torda 1975). However, another less direct, but more objective, way to test the perception of efferent

---

* Though more controversial, the impressions of movement of a 'phantom limb' in amputated patients are relevant to this point. In an editorial appearing in 1875 (unsigned, but attributed to H. Jackson), there was a question as to whether certain of the patients could have a very precise sensation of the movements or positions of the phantom hand. Patients stated: 'My hand is now open, it is closed, . . . I touch the thumb with the little finger . . . It is now in the writing position'. Jackson concluded that 'the volition to move certain parts is accompanied by a mental condition which represents the quantity of movement, its force, an idea of a change in position of these parts in the consciousness', in Psychology of the nervous system, *British Medical Journal*, 1 October 1875, p. 462). It was shown later that sensations of phantom limb movements were accompanied by muscle contraction in the stump. If these contractions were blocked, the sensations of movement disappeared (Henderson and Smyth 1948).

activity is to ask subjects to indicate the *quantity of effort* that they have to put into achieving a given task. Gandevia and McCloskey (1977) asked subjects to press a lever with one thumb in order to produce a reference tension, displayed visually on an oscilloscope screen. With the other thumb, they had to press another lever so as to match, without visual control, the muscular contraction or effort produced by the 'reference' thumb. During partial curarization on the side of the reference thumb, the subjects could still produce the reference tension by pressing the lever, but with the other thumb they indicated a much larger muscular effort than normally required to produce the same tension (Fig. 5.4).

Gandevia and McCloskey (1977) also used the perceived heaviness of weights as a measure of muscular effort. The same matching technique as above was used, i.e. subjects chose weights with one arm until the heaviness perceived with that arm matched the reference weight lifted by the other arm. During partial paralysis, the subjects chose exaggerated weights, hence indicating an increase of the heaviness perceived with the weakened arm. The same result was obtained with muscular fatigue instead of partial paralysis. McCloskey *et al.* (1974) showed that after the reference arm had supported the weight for some minutes and had become fatigued, weights heavier than the reference weight were chosen to match it (Fig. 5.4).

There are several possible ways of interpreting the sensations reported by paralysed subjects on their attempts to move, as well as their perception of increased heaviness as measured psychophysically. Gandevia (1982) suggested that neural traffic in motor corticofugal paths might be read and used as the relevant signal for the observed illusions. Indeed, complete paralysis following lesions of the motor pathway at the cortical level is not accompanied by sensations of increased effort or heaviness, precisely because no traffic occurs in the motor pathways after such lesions. Sensations reappear during partial recovery of movements, when neural traffic is re-established. The same hypothesis would account for permanence of sensations of effort in all cases of distal paralysis, where corticofugal neural pathways are not altered. The mechanism proposed by Gandevia (1982) for explaining perception of motor commands is therefore very similar to that postulated in the 'corollary discharge' or 'efference copy' models, which postulate that part of the motor output is sampled at some point of the motor pathways, and sent to sensory analysers (see above, and Chapter 4). The same basic principle was exploited in a rather extreme way by Granit (1972). Granit, who opposed the notion of sensation of innervation (which he considered as a 'physiological dead end'), made the remark that the linkage of alpha and gamma motoneurons during voluntary movement in fact represented the basis for a 'corollarization' of motor output. Some aspects of this idea are discussed in the next section.

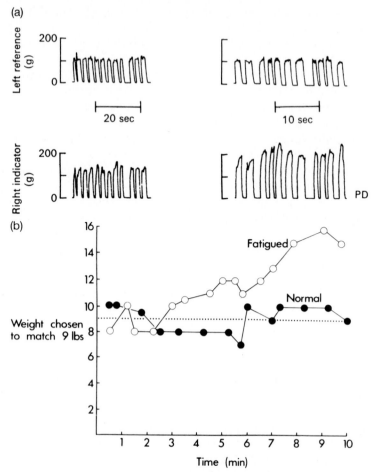

Fig. 5.4 Perception of quantity of effort by normal subjects.

(a) Matching a reference isometric contraction produced by extension of one index finger, with an extension of the other index. Left, control situation. The subject extends his left (reference) index finger on successive trials to reach a visually displayed reference tension (upper row). He reproduces blindly the same extension with his right (indicator) index finger. Right, partial curarization of the left arm. The subject is still able to produce the reference extension (upper right) with his left index finger; however, he indicates an increased effort to match the reference extension. (From Gandevia and McCloskey 1977, with permission.)

(b) Matching a weight supported by one arm, by choosing an apparently equal weight with the other arm. Subject was given a 9 lb weight to support by contraction of biceps brachialis of one arm (reference arm). With the other arm, he chose apparently equal weights. When the reference arm was rested between successive trials, the subject chose weights close to the reference (black circles). When the reference arm supported its weight continuously, it became fatigued and weights heavier than reference weight were chosen (open circles). (From McCloskey *et al.* 1974, with permission.)

5.3.3. The collaboration of afferent and efferent mechanisms

By comparison with the role of proprioceptive mechanisms, the contribution of efferent mechanisms to limb position sense appears very modest. As a matter of fact, there is no experimental or introspective evidence that sensations generated by motor commands may be used for monitoring the position of a limb—only the amount of effort to perform a movement or to maintain a position seem to be perceived.

It might be useful, however, in order to understand the contribution of efferent mechanisms, to return to the original definition of the efference copy. In von Holst's (von Holst and Mittelstaedt 1950) formulation, the role of the efference copy is to cancel undesirable sensory messages resulting from self-generated movements. This function does seem to be demonstrable in the motor system, in the sense that the gamma motoneuron discharge (which may well be considered as a 'copy' of the alpha command) acts to cancel the effects of the centrally programmed muscular shortening on the muscle spindles. Miles and Evarts (1979) therefore assumed that 'the control process utilizing alpha–gamma co-activation may be viewed as one in which efference copy allows spindle afferent discharge to signal the *distorsion* or *error* of a centrally-programmed movement' (p. 358).

A somewhat different version of the same concept has been offered by McCloskey *et al.* (1983a). They postulated that the efference copy signal derived from the gamma command would have the function of offsetting the fusimotor induced spindle activity. Consequently, only the spindle activity resulting from the shortening of the muscle by the alpha command would constitute the kinaesthetic signal. This hypothesis would thus attribute to the efference copy a purely ancillary function in position sense. This function would be consistent with the observed lack of movement illusion when the efference copy is monitored in isolation from its natural context.

Finally, another hybrid conception of position sense has been developed by Feldman and Latash (1982a). These authors started from the fact that there exists an invariant dependence between muscle force and length (Feldman 1966). Each value of the central command therefore defines a point of equilibrium of the agonist and antagonist muscle-load ensembles, which corresponds to a stationary joint position (the equilibrium point hypothesis, see Chapter 1). The muscle length related to the equilibrium point is identified by the central nervous system through the proprioceptors. The two variables of muscle force or tension and muscle length define a 'sensomotor space' which, in Feldman and Latash terms, represents a basis for integration of sense of effort (as known from efferent copy signals) and sense of position (as known from proprioceptive coding of muscle length). Proprioceptive signals alone cannot give the proper information as to joint angle for the reason stated above, i.e. that signals arising from the spindles include not only muscle length-related activity, but also fusimotor activity, itself related to the amount of effort needed to maintain the position. Efference copy from the central commands reaching the agonist and antagonist muscles would therefore be necessary for *calibrating* proprioceptive activity. According to Feldman and Latash (1982b), kinaesthetic illusions may arise from distortions in the central evaluation of the afferent/efferent components of joint position information.

5.4. The role of position sense information in the mechanism of proactive control of movements

Position sense provides information related to joint angles. This information is necessary but not sufficient to drive the hand at targets located outside the body space. Hand position must in addition be calibrated within a 'working space' that defines the relationships between body, hand and target positions. Vision seems to play an essential role in this mechanism.

Until recently, visual control of movement was considered as relying on feedback mechanisms, by which visually detected errors could be corrected during execution and improve accuracy. This mechanism, which has been fully described in Chapter 3, has strong limitations due to the relative slowness of the visuomotor loop used for corrections based on visual feedback. In this section another mechanism for visual control of movements is proposed, whereby visual input acts proactively, in conjunction with limb position sense, to direct the limb at the target and to achieve co-ordination between the limb segments during target acquisition. In addition, it will be shown that visual information involved in this mechanism cannot be dissociated from information about the position of the eye during visual fixation of the target.

5.4.1. Vision of the hand prior to movement: visual and proprioceptive maps

The main experimental evidence supporting a mechanism for movement accuracy, involving simultaneously vision and position sense, comes from the findings of Prablanc and his colleages. In their experiment (Prablanc *et al.* 1979b), subjects had to point at visual targets in the same experimental set-up as described in Chapter 3. Pointing was made in the 'no visual feedback' condition, i.e. no visual information was available from the hand during the movement. The experimental variable was the amount of visual information about limb position available *prior to* the movement. In one condition, no information was given, and the limb remained invisible throughout the trial. In the other condition, the hand was visible at its initial resting position, and the light was turned off at the onset of the

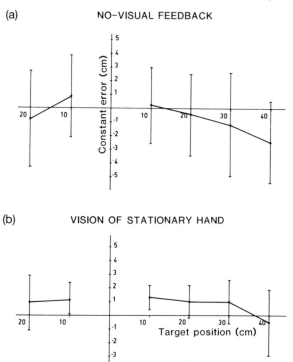

(a) NO-VISUAL FEEDBACK

(b) VISION OF STATIONARY HAND

Fig. 5.5 Effect of vision of the stationary hand prior to a movement, on reaching accuracy.

(a) Subjects reached, with their right hand, targets located 10, 20, 30, 40 cm from midline to the right and 10, 20 cm to the left. Head was fixed, eye movements were free. In the absence of visual feedback note constant error and large scatter of hand pointings (same figure as Fig. 3.9b).

(b) Same condition except that the hand was visible until it moved. Note reduction of both constant and variable errors. (Replotted from Prablanc *et al.* 1979b.)

pointing movement. The two conditions were therefore characterized respectively by presence or absence of visual information about the static position of the hand relative to the body and to the target.

Pointing errors were measured in both conditions. In the condition with no vision of the hand, large constant and variable errors were recorded, in conformity to the results previously described in the 'no visual feedback' condition (Fig. 5.5a). By contrast, in the condition of vision of the stationary hand before the onset of the movement, the pointing performance was clearly improved, at least for target eccentricities up to 30° from midline, due to reduction of both the constant error and the variable error (Fig. 5.5b). This result therefore demonstrates that proprioceptive and central signals for hand position sense (the 'proprioceptive map') are unsufficient by themselves to efficiently encode the position of the hand within the working space. In other words, *position sense must be calibrated by vision in order for the proprioceptive map that encodes limb position relative to the body, and for the visual map that encodes the position of the limb relative to the target, to be matched to each other* (see Jeannerod and Prablanc 1983). Systematic disconnection between these two maps at the central level might represent a possible explanation for pathological disorientation resulting from cortical lesion. This point will be developed in Chapter 6.

There are very few data available as to how often the proprioceptive map needs to be updated in order to stay in register with the visual map. The key experiment to do would be, using the Prablanc *et al.* (1979a,b) paradigm, to expose the hand prior to the movement only on the first of a series of trials, and to measure on how many of the subsequent trials with the hand not exposed the effect of visual exposure lasted. In the absence of such data, one can only speculate that without vision, position sense will rapidly loose its relationship to extrapersonal space. An experiment by Thomson (1980) seems to confirm this view. Subjects inspected for 5 s targets located 1–20 m from them. After being blindfolded, they were requested to walk immediately toward the target, and to stop walking at its estimated location. The subjects were quite accurate for targets at distances up to about 10 m. Beyond this distance, large localization errors were observed. Thomson's conclusion was that the visual representation of the target could not be stored for more than about 8 s (the time that it took the subjects to walk 10 m). Indeed, if subjects had to wait before walking to the target, the distance that they could reach accurately was reduced. This finding is thus consistent with the idea of a progressive drift of the two maps with respect to each other, when vision of the relative positions of the body and the target is precluded.

5.4.2. Foveation

In order to explain the less marked effect of vision of the stationary hand on accuracy of pointing at the most eccentric targets (beyond 30°, Fig. 5.5(b)), Prablanc *et al.* (1979b) compared for each trial the latency of the hand movement with the time taken by the eye to reach the target. They found that in movements directed at targets within 30°, the eye reached the target and foveation was achieved before the hand began to move, that is, while the hand was still visible. By contrast, in movements directed at the 40° target the time for the eye to reach the target was longer than the hand latency, so that foveation could not be completed before the hand moved and disappeared (Fig. 5.6). The fact that accuracy deteriorated in the second condition indicates that fixation of the gaze on the target

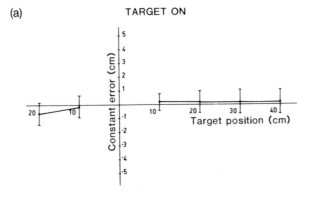

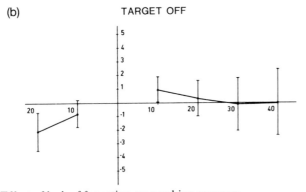

Fig. 5.6   Effect of lack of foveation on reaching accuracy.

(a) Subjects reached targets with their right hand. Normal visual feedback condition (same figure as Fig. 3.9a).

(b) The targets were turned off at the onset of saccades, so that the eye movements ended in the target area but could not foveate the targets. Note increase in variable error of hand pointings. (Replotted from Prablanc *et al.* 1979a.)

(foveation) allows a better encoding of the initial hand position relative to the target, provided the hand and the target can be seen simultaneously before the onset of the hand movement. The importance of foveation for movement accuracy also explains the result of another experiment of Prablanc *et al.* (1979a), where the targets to which eye and hand movements were directed were turned off at the onset of the eye movement. Although the eye landed very close to target position, the hand accuracy was severely impaired (Fig. 5.6b). This result can thus be interpreted as a consequence of the lack or the incompleteness of foveation.

In order to analyse in detail the nature of visual signals operating during foveation, Prablanc *et al.* (1986) undertook another study where visual information relating to the target was manipulated. In these experiments, the targets (20, 30, 40, and 50° from the midline on the side ipsilateral to the pointing hand) were turned off at precise times during the reaching movement (Fig. 5.7, left). Pointing was made in the 'no visual feedback' condition in all trials, so that visual feedback during the movement could not be suspected of interfering with the results. However, vision of the hand was available prior to pointing. Pointing accuracy was clearly a function of the time during which the target was shown. When the target was turned off at the onset of the hand movement, large constant errors were recorded for the 30, 40, and 50° targets. It should be noted that during pointing at these targets, the eye movement was not completed at the time of target offset and foveation was not achieved, whereas during pointing at the 20° target, foveation was achieved before the target disappeared (Fig. 5.7, right). These results clearly confirm that foveation provides additional information for guiding the hand at the target.

The experiments reported in the above two sections (4.1 and 4.2) seriously question the classical interpretation of reaching movements as composed of large initial ballistic phases and short terminal feedback-controlled phases (see also Chapter 3). Instead, the results indicate that visual information is needed prior to and throughout the movement to achieve its full accuracy, and that a large amount of this information is *not* of the feedback type. The role of foveation in this mechanism seems critical. During foveation the gaze axis is aligned with the position of the target in space. Provided the hand is also visible at its initial position, foveation will result in aligning the origins of the visual and proprioceptive co-ordinate systems or maps on the target position.

5.4.3. Role of the representation of target position

The demonstration that on-going visually directed movements can be controlled in the absence of visual feedback implies a mechanism of comparison between visual/oculomotor signals about the target position and extra-visual signals for hand position. The controlling mechanism can be conceptualized as a three-level mechanism (Jeannerod and Prablanc 1983). One level would correspond to the visual 'map', where target position is accurately represented; the other level would be a comparator where the error between the internal representation of target position and the limb position would be monitored. Error signals generated by the comparator would account for on-line correction of limb position. Finally, the third level would correspond to the classical visual feedback mechanism for correction of terminal errors.

In order to demonstrate this possibility, Pélisson *et al.* (1986) undertook an experiment similar to those of Prablanc *et al.* (1979a and 1986), except that targets occasionally made double steps. The subjects' task consisted in pointing at the targets with their invisible hand. The initial target steps were from midline to 20, 30, 40, and 50°. The second steps were of a

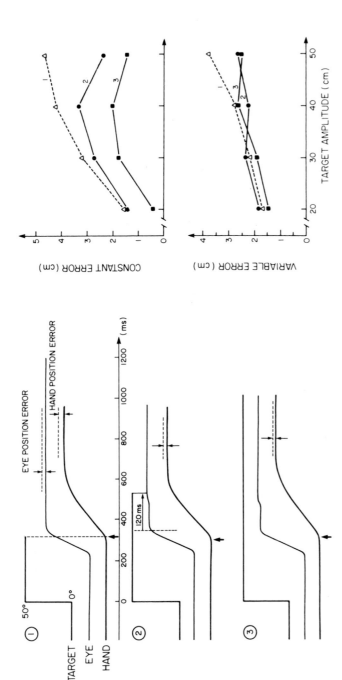

Fig. 5.7   The role of foveation in the accuracy of reaching. Constant error (CE) and variable error (VE) of the hand have been plotted as a function of target eccentricity in three different experimental conditions. In condition 1 (upper left) the target was turned off before completion of the eye movement, so that no foveation occurred. Indeed, it is in this condition that errors were the largest. In condition 2 (middle left), the target was turned off after completion of foveation but before the end of the reaching movement. This procedure improved accuracy with respect to the control condition (condition 3, target permanently lit, lower left). (From Prablanc *et al.* 1986, with permission.)

smaller amplitude (e.g. from 30 to 32°, 40 to 44°, 50 to 54°) that is, sufficiently large to make possible pointing corrections clearly visible. The second steps, when present, were triggered at maximum saccadic velocity, a procedure that turned out to be quite advantageous for the experiment: subjects were not aware of the target jumps and were never able to report the occurrence of a double step. The results showed that subjects consistently undershot target position in both the single-step and the double-step trials, due to the lack of terminal visual feedback. Nevertheless, the distributions of pointing positions for double-step trials were significantly shifted with respect to those of the corresponding single-step trials (Fig. 5.8(a)), indicating that the subjects corrected their hand trajectories in order to reach the final target positions. This result has important implications. If one assumes that final target position was not properly encoded until foveation occurred, then the hand had to wait until the saccade was completed, before corrections needed to reach the displaced target could be generated. Consequently, these corrections necessarily took place *after* the hand movement had begun.

One could hypothesize that corrections were in fact new movements resulting from reprogramming. Pélisson *et al.* (1986) clearly showed that this was not the case. Indeed, the durations of pointing movements toward displaced targets followed the same linear relationship with amplitude as for movements toward stationary targets (Fig. 5.8(b)). In other words, the increased duration on double-step trials reflected only the additional distance that the hand had to move, indicating that there was no reprogramming of movements to accommodate the second target displacement. Another argument as to this point was given by the kinematic analysis of the pointing movements. Indeed, if reprogramming occurred, it should become visible as a secondary movement and as a re-acceleration of the trajectory. No such re-accelerations were seen for the double-step trial movements, which therefore, did not differ kinematically from the single-step trial movements.

The fact that in the present case, corrections were generated on the feedforward mode had an obvious advantage. Because the peripheral feedback loops were by-passed, and no reprogramming occurred, corrections took little extra time and were compatible with rapid adjustment of the responses. The lack of subjective awareness about target displacements and the corresponding corrections, also can be interpreted as contributing to the same time saving process. It would be interesting to know, by triggering the target step later and later in the movement sequence, at which point secondary movements would begin to be generated, that is, at which point subjects would shift to the other, feedback mode, of correction with reprogramming.

There are other examples of fast corrections in ongoing movements. Abbs and his colleagues have described rapid compensation of pertur-

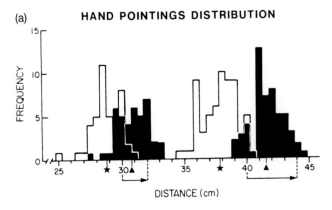

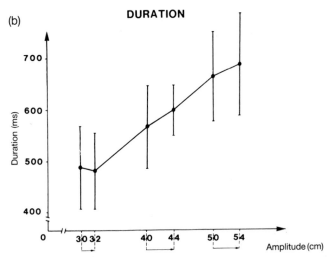

Fig. 5.8 (a) Distribution of pointings directed at stationary and displaced targets. For a single step to 30 cm, the distribution of pointings represented by the white histogram (mean indicated by the star) undershoots the target. When the target is further displaced from 30 to 32 cm (double step), the distribution of pointings represented by the dark histogram (mean indicated by the filled triangle) undershoots the 32 cm target. Notice, however, that the distance between the star and the triangle is equal to the target shift, i.e. subjects fully correct their ongoing response. The same applies for a target shift from 40 to 44 cm, and for a target shift from 50 to 54 cm (not shown).

(b) Duration of hand pointing movement (mean and standard deviations) versus amplitude of the target final position. When a target is displaced for instance from 40 to 44 cm, the duration increases as compared to responses to the 40 cm single step, but the total duration corresponds (by extrapolating from the observed durations to 40 and 50 stationary targets) to the predicted duration of pointings to a stationary target appearing at 44 cm. Thus, no additional reaction time or extra processing time appears within the total duration to the double step stimulation, nor is it associated with a higher variability of movement duration. (From Pélisson *et al.* 1986.)

bations applied to articulators during speech. Abbs and Gracco (1984) reported an experiment where the lower lip was unexpectedly pulled down during production of the phoneme /ba/ by normal subjects. They showed that lip closure was nevertheless achieved and that the phoneme was correctly pronounced in spite of the perturbation (Fig. 5.9(a)). Such a compensation implied that lip closure was performed by a larger lowering of the upper lip, and the perturbation was compensated within a delay compatible with the production of the phoneme. Indeed, EMG activity of orofacial muscles was found to increase as a consequence of the perturbation (Fig. 5.9(b)). For the orbicularis oris superior, responsible for upper lip depression, the latency of EMG activation was 22–55 ms, although for the o. oris inferior (the lower lip elevator), EMG activity increased several tens of milliseconds later. The short term increase in EMG activity of the o. oris superior accounted for the close-to-normal lip closure.

Abbs and Gracco (1984) concluded that activation of these two independent muscles reflected two different compensatory mechanisms. Late activation of the lower lip elevator reflected reprogramming triggered by proprioceptive feedback. Early activation of the upper lip depressor, by contrast, reflected an open-loop adjustment, that is, independent of sensory feedback in the usual sense, but nevertheless relying on proprioceptive signals generated by the ongoing movement. Such a conception of movement 'correction' would clearly fit into the definition of feed-forward mechanisms given by Arbib (1981): 'A strategy whereby a controller directly monitors disturbances to a system and immediately applies compensatory signals to the controlled system, rather than waiting for feedback on how the disturbances have affected the system' (p. 1466). This definition implies that the system must have access to a representation of the goal of the ongoing movement and can distribute the commands to the muscles in such a way that the goal can be reached (see below).

Another illustration of this mechanism is given by the rapid compensation of goal-oriented eye movements during perturbations applied to the head. Pélisson and Prablanc (1986) requested subjects to track targets appearing in their peripheral visual field, during unexpected passive head displacements directed either in the same direction as the targets, or in the opposite direction. The design of the experiment was such that neck proprioceptive and vestibular signals as to direction and velocity of the head displacement were available to the oculomotor system a brief period time (about 200 ms) prior to saccade initiation, and during the saccade itself. Saccades generated in these perturbed conditions were found to be as accurate as saccades generated with the head stationary. This implied a subtle correction. Saccades generated against or with the head movement had peak velocities respectively higher or lower than saccades generated in normal conditions. At the same time, the duration of these saccades had to be changed in order to adjust the velocity change, and to avoid overshooting

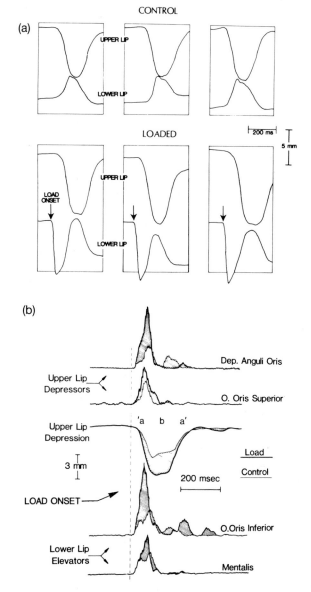

Fig. 5.9 (a) Rapid compensation of perturbations during speech movements.
Movements of upper and lower lips during lip closure for producing the phoneme /ba/. Upper row, normal trials. Lower row, perturbed trials. Perturbation (a pull of the jaw downward) is indicated by arrows.

(b) Changes in upper lip depression and in EMG activity of perioral muscles following a downward pull of the jaw. Note increase in lowering of the upper lip and increase in EMG activity of the upper lip depressors. Vertical dotted line: load onset. (From Abbs and Gracco 1984, with permission.)

or under-shooting the target. Thus the proper adjustments occurred in much less than one reaction time.

Similarly, Laurutis and Robinson (1986) showed that brisk perturbations applied to the head during a target-oriented saccade were compensated by an alteration in saccade duration, so that the accuracy was preserved. Their interpretation was that gaze (eye + head) position was automatically compared with the desired gaze position and that an amount of movement corresponding to the amplitude of the head displacement was added to, or subtracted from, the saccade during its execution.

These data stressing rapid alteration of ongoing movements (see also Megaw 1974; Cooke and Diggles 1984) suggest an on-line control of execution based on a representation of the goal of the movement. This mechanism can be conceptualized as a continuous comparison between the represented goal and the instantaneous state and position of the effector, what Abbs (1982) has called an 'afferent-dependent feed-forward process'. Such a process implies that neural information about target position (in visual space for pointing movements, in phonemic space for speech movements), and information about state and position of the effector, can be translated into a common language, understandable by the comparator. At present, there is no way, other than pure speculation, of conceiving this mechanism in neurological terms.

5.4.4. The role of position sense in visuomotor control. Pathological observations

One way of turning around this conceptual difficulty is to use pathological situations where information about target position and about limb position are disconnected. Lesions of somatosensory pathways in humans offer good examples of such a disconnection.

In this section, the observations of two patients are reported to demonstrate the effects of loss of position sense on visuomotor behaviour. The two patients differed by the level at which information for position sense was impaired (at the brainstem level in one case, at the cortical level in the other).

(a) Lesion of the somatosensory pathways at the brainstem level

Patient Tah was a 37-year-old man who suffered a severe head injury with a fracture of the occipital bone. CT scan (Fig. 5.10) revealed the presence of a bone fragment protruding in the right anterolateral quadrant of the medulla. This fragment was likely to have damaged the lemniscal sensory pathways controlling the ipsilateral limbs. When he was first examined, the patient presented a mild hemiparesis of right upper and lower limbs, and a complete loss of stereognosis and position sense on the same side. The loss of tactile sensation was less complete. Six months after the trauma, i.e.

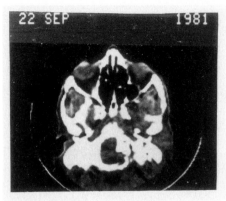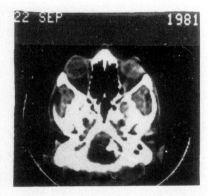

Fig. 5.10   CT scan of the base of the skull in patient Tah. Note bone splinter protruding into the occipital canal on the right side. (From Mauguière *et al*. 1983, with permission.)

Table 5.1. Averaged movement parameters in patient *Tah*.

	Normal hand	Affected hand	
	No visual feedback	Visual feedback	No visual feedback
Duration	560	925 (105.6)	1096 (188.8)
Time to velocity peak	300	370 (63.2)	400 (101.9)
Maximum grip size	54.0 (1.7)	51.0 (5.4)	74.3 (5.6)
Final grip size	46.0 (3.6)	34.7 (6.5)	62.7 (13.1)

Duration and time to velocity peak are in ms; maximum and final grip size (before contact) are in mm. Numbers in parenthesis are SDs.

when the tests reported here were performed, the hemiparesis had completely cleared, although the sensory deficit remained unchanged. A specific component of the somatosensory-evoked potential in response to stimulation of the right hand was permanently abolished (for a complete description, see Mauguière *et al*. 1983).

Prehension movements were examined in this patient by using the technique described in Chapter 2. Movements performed by the patient with his hand contralateral to the lesion (the 'normal', left hand) were normal in every respect. Mean values of maximum grip size and final grip size before contact for that hand in the 'no visual feedback' condition, are given in Table 5.1. Movements performed by the hand ipsilateral to the lesion (the 'affected', right hand) differed whether visual feedback was available or not during the movements. In the 'visual feedback' condition, duration of the movement was found to be longer than with the normal hand (Table 5.1). Grip size, however, was correctly adapted to the size of the objects (Table 5.1), and finger posturing was such that accurate grasps were

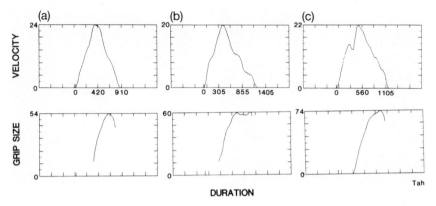

Fig. 5.11   Change in velocity of the transportation component and grip size of the manipulation component as a function of movement duration in patient Tah. (a) Affected hand, 'visual feedback' condition. (b, c) Affected hand, 'no visual feedback' condition. Movement represented in (b) is the first movement recorded in the 'no visual feedback' condition. Maximum grip size: (a), 54.5 mm; (b) 66 mm; (c) 73.7 mm. (From Jeannerod 1986a.)

performed (Figs 5.11(a), 5.12(a)). The normal character of finger grip formation with the affected hand was obviously critical for assessing the integrity of the pyramidal tract in this patient. By contrast, in the 'no visual feedback' condition prehension movements were greatly altered. Not only was movement duration exaggeratedly long (Table 5.1), but also finger grip was either absent or incomplete. The first movement recorded in the 'no visual feedback' condition showed a complete lack of grip formation (Figs 5.11(b), 5.12(b)). In subsequent movements incorrect grip formation was observed, with exaggerated opening of the index and third fingers and incomplete finger closure (Table 5.1, Figs 5.11(c), 5.12(c)).

The main problem of this patient was thus to achieve a motor pattern in the absence of a direct visual control of his affected hand. In other words, the visual signals relating to the shape of the object to be grasped could not be used to build the correct motor commands for the fingers. Position sense is an essential aspect of visually goal-directed actions, perhaps in the same way as visual signals are an essential aspect of the control of the positions of body parts.

(b) Loss of position sense following a parietal lesion

Similar sensorimotor deficits were also observed by Jeannerod *et al.* (1984) in their patient R.S. with sensory loss following a parietal lesion. This patient was a 52-year-old woman who presented with an infarction of a large zone of the parietal lobe following occlusion of the posterior parietal artery, a branch of the sylvian artery (for a complete description, see Jeannerod *et al.* 1984). The CT scan revealed a clear-cut hypodensity of the

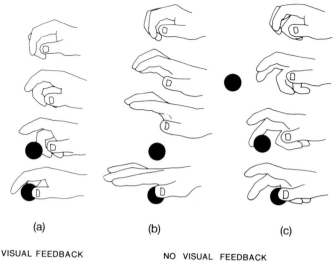

(a)	(b)	(c)

VISUAL FEEDBACK   NO VISUAL FEEDBACK

Tah

Fig. 5.12 Pattern of finger grip during reaching in patient Tah. (a) Affected hand, 'visual feedback' condition. (b, c) Affected hand, 'no visual feedback' condition. Movement represented in (b) is the first movement recorded in the 'no visual feedback' condition. Redrawn from film. (From Jeannerod 1986a.)

whole post-central gyrus, except for its mesial part, and of the supramarginal gyrus. The pre-central gyrus was apparently spared. The thalamus seemed to be intact, although the thalamic parietal radiations were possibly destroyed (Fig. 5.13).

Although deafferentation in this patient was not of peripheral origin, her behaviour was strikingly similar to that observed in patients with peripheral deafferentations. Sensory loss was virtually complete for the right hand and wrist. At the level of the right arm and shoulder, tactile anaesthesia was less severe, since strong stimuli were detected and grossly localized. Sensibility to cold, warm, and vibratory stimuli was impaired with the same distribution as for the tactile stimuli. Sensations evoked by passive movements were abolished at the right fingers and wrist. At the right elbow the direction of passively induced movements could not be consistently detected. In the absence of visual control, R.S. made frequent errors whenever she indicated the direction of passive movements verbally or tried to match the angle of her right elbow with that of her left. The level of detection was influenced by the velocity of the movement. Better detection was achieved when movements were applied briskly, whereas slow movements were never detected. Finally, at the right shoulder, the direction of passive movements could be detected with less error and matching with the other arm was reasonably good.

R.S. did not at first use her right hand spontaneously, but later she used it in everyday life for many types of actions, provided she could control her

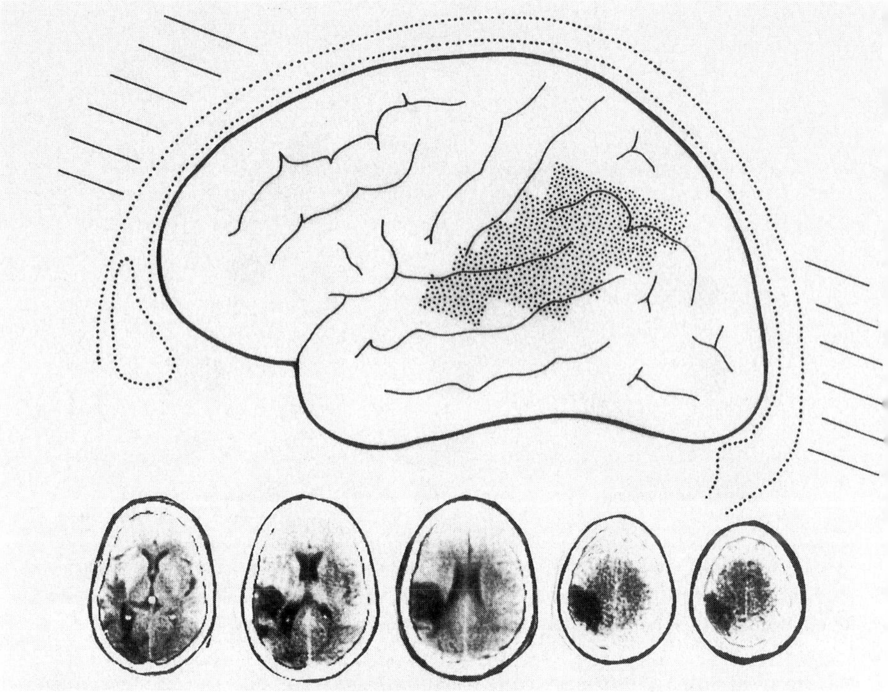

Fig. 5.13 Reconstruction of the lesion of R.S. The extent of the destroyed area on five CT scan sections has been outlined on a lateral view of the left hemisphere. (There is no section available above the fifth slice). (From Jeannerod *et al.* 1984.)

movements visually. Without visual control, movements with that limb were awkward and inefficient. This dependence of movement accuracy on visual feedback became clearly apparent when R.S. was tested for her ability to make sequential distal gestures, such as drumming or tapping with the fingers, under visual control and without vision. In the absence of vision, the rate of tapping with the right index finger could not be sustained and rapidly degraded. However, an almost normal tapping rate could be sustained when the taps were made audible. Similarly, R.S. was able, under visual control, to touch successively with the tip of her right thumb the tip of the four other fingers of the right hand. Without vision, however, she was unable to do so. In her attempts, movements were clumsy and spatially disoriented. Finally, under visual control, she was able to draw figures or letters in the air with the fingers of her right hand. Without vision, the same movements could not be executed. In her attempts, R.S. described verbally what she thought she was doing but her fingers appeared to move randomly and with a very limited amplitude.

Prehension movements executed by this patient with her arm ipsilateral to the lesion were normal in every respect for what concerned both the

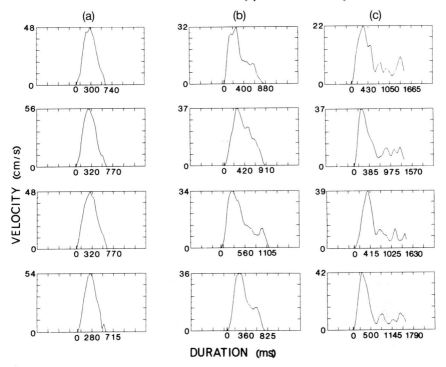

Fig. 5.14 Change in velocity of the transportation component as a function of movement duration in patient R.S. (a) normal hand, 'no visual feedback' condition. (b, c) Affected hand in the 'visual feedback' and 'no visual feedback' conditions, respectively. Four movements are shown in each condition. (From Jeannerod 1986c.)

transportation component and grip formation (Figs 5.14(a) and 5.15(a)). With her hand contralateral to the lesion, prehension was influenced by whether visual feedback from the moving hand was available or not. In the 'visual feedback' condition the transportation component of prehension appeared to last longer than with the normal hand, due to the occurrence of secondary velocity peaks during the deceleration phase (Fig. 5.14(b)). Grip formation was inaccurate and resulted in undifferentiated grasps with the palmar surface of the whole hand (Fig. 5.15(b)). In the 'no visual feedback' condition, only the initial part of the transportation component was normal. Following the first velocity peak, the hand wandered above the object location without achieving the grasp (Fig. 5.14(c)). No sign of grip formation could be observed (Fig. 5.15(c)).

These two clinical cases represent good illustrations of the necessary interaction between visual and somatosensory signals in controlling visually goal-directed actions. The first patient, although his lesion was restricted to the medullar somatosensory pathways, was nevertheless

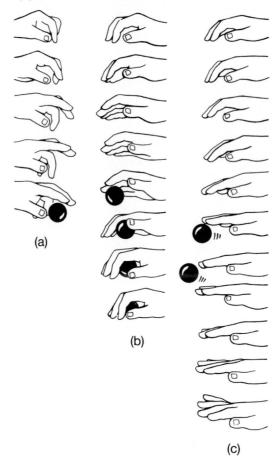

Fig. 5.15 Pattern of finger grip during reaching in patient R.S. (a) normal hand, 'no visual feedback' condition. (b, c) Affected hand, in the 'visual feedback' and 'no visual feedback' conditions, respectively. Redrawn from film. (From Jeannerod *et al.* 1984.)

unable, in the absence of direct visual control, to correctly adapt his fingers on the affected side to the visual shape of the objects. Because the processing of visual information concerning object shape obviously had not been affected by such a peripheral lesion, one has to assume that the defect in grip formation reflected the impossibility of visual signals being integrated with somatosensory signals concerning finger posture and movements. In addition, the fact that interruption of the somatosensory information flow at the cortical level produced the same effects on grip formation as interruption at the peripheral level (as shown by the observation of the second patient) indicates that the part of that information that transfers through the cortex is critical for this mechanism. This point will be developed in Chapter 6.

6. Impairments in visuomotor control following cortical lesions

In the previous chapters, several levels of organization of visuomotor control were examined, namely the respective contributions of central (programming) versus peripheral (feedback) mechanisms; the role of eye-in-head or head-on-body positions as references for controlling direction and accuracy of the movements; and the relative independence of input–output channels for the control of proximal and distal segments of the musculature. In the present chapter, effects of localized cortical lesions on these various aspects of reaching behaviour will be reported from clinical cases. Although this attitude implies a strong inference about structure–function relationships, it will find its justification, first in the results of animal experiments where specific impairments in visually-guided behaviour have been related to well-defined lesions, and second, in a model that attempts to reconcile the effects of lesions in clinical patients and the behavioural studies in normal subjects.

The problem, in the clinical study of reaching impairments, is to identify a pathology specifically related to sensorimotor (or visuomotor) co-ordination, i.e. related to the relation between the acting body and its environment. Such reaching disorders were first identified under the heading of visual 'disorientation', a terminology implying both a preservation of visual processes responsible for the perception of objects and a difficulty for the subject to orient his action toward them. This definition takes account of the fact that spatially directed actions not only rely upon intrinsic visual cues like size, relative position, or relative motion of objects ('allocentric' cues), but also require the availability of 'egocentric' cues, related to the position of the body with respect to the surrounding objects.

Although visual disorientation was at an early date identified as a clinical sympton (Balint 1909; Holmes 1918), it has been systematically reinvestigated in recent years. One reason for this revival of interest is that disorders in reaching have been recognized as belonging to a broader entity involving both behavioural and cognitive impairments, and specifically related to lesions located within the right cerebral hemisphere. The enormous prevalence of language in humans may have exaggeratedly drawn the attention of clinicians to the effects of lesions of the left hemisphere, and, by contrast, obscured the effects of lesions of the right, silent, hemisphere. This may explain why these effects were not fully recognized until recently.

6.1. Cognitive and behavioural disorientation following cerebral lesion: a historical account

A detailed history of the clinical studies relating spatial disability to specific cortical or hemispheric lesions has been made recently by Benton (1982). Only the major landmarks will be recalled here.

6.1.1. R. Balint

Historically, the first important paper on visual disorientation was that of R. Balint (1909), who described a single patient in whom he had identified a symptomatic triad: psychic paralysis of gaze; optic ataxia (optische Ataxie); and spatial disorder of attention. Although this patient had normal eye movements and no visual field defect, he would not orient his gaze to the left. According to the author, the patient could see objects, but did not look at them, which justified, he thought, the terminology of 'psychic paralysis of gaze movement', already used in the context of limb apraxia. The resulting attentional deficit was predominant in the left half of space. It was also characterized by the fact that the patient could see only one object at a time.

When asked to reach for objects, the patient behaved differently with either hand. Although movements with the left hand were normal and easily reached the target, those with the right hand erred in all directions until the patient eventually bumped his hand into the object. Balint had noted that disorientation of movement with the right hand was due to a problem with visual control for that hand. If the patient first pointed to the object with his left hand, then his right hand could also reach accurately. Position sense of the right hand was preserved, because the patient could reproduce with that hand positions imposed on the left hand. Balint concluded that this inability to reach under visual control resembled tabetic ataxia, where patients cannot use their limbs under proprioceptive control alone, hence the term 'optic ataxia'.

Post-mortem examination of the Balint case revealed that the main lesion was located in the posterior parietal areas, involving the angular gyrus and the anterior part of the occipital lobe on both sides.

6.1.2. G. Holmes

Holmes contributed to the problem of visual disorientation by reporting on several cases of gunshot injuries in young patients. Their common disturbance was characterized by the author as a loss of 'the power of localizing the position in space and the distance of objects by sight alone'.

Six cases were reported in the paper by Holmes (1918). All six patients had a homonymous visual field defect, in the right visual field in two cases, in the left visual field in three cases. In the last case, the loss of vision affected the lower two quadrants. Visual acuity, however, was clearly preserved, although some of the patients complained of 'misty' vision during the early stages of their lesion. Visual objects were recognized in all cases. Visual memory was normal in most of them.

Eye movements were abnormal. Although there was no ocular palsy, patients were frequently found to be unable to move their eyes on command. In attempting to fixate objects, the eyes tended to roll in all directions until they fell on the object as if by chance. When fixation was finally achieved, the patients had difficulties in maintaining it if the object was moved around. Visual observation of a large surface was poor. In addition, convergence and accommodation were found to be frequently impaired. Holmes considered these oculomotor defects as central to the understanding of visual disorientation. This point will be further documented and discussed in a special section (see below).

The disturbance in visual disorientation (Holmes did not use the term 'optic ataxia' coined by Balint), very similar in the six reported cases represented the most striking symptom. Typically, the patient of Case 1, when asked to take or touch an object placed in front of him, would often project his arm in a totally wrong direction: 'when he failed to touch the object at once he continued groping for it until his arm or hand came into contact with it, in a manner more or less like a man searching for a small object in the dark'. This difficulty in localizing objects in space was greater when the object lay outside his central visual field. A common complaint of the subject was that he could see the object, though he was not sure where it was. When walking, he had great difficulty finding his way around an obstacle placed in his path, or to an object placed at some distance. A second patient had the same problems localizing objects placed outside his line of sight, especially at a distance. He would in all cases stretch out his arm and search for objects, as if they were always located at arm's length. He was also unable to find his way back to his bed if any obstacle had been placed in his path. The patient of Case 3 suffered with 'an extremely gross disturbance of localization in space by vision': he groped for objects, brought his hand beyond them, and made errors in all directions. His errors were always greater when the objects he attempted to touch were outside central vision. In some patients, the reaching errors were observed exclusively in the sagittal dimension. Finally, Holmes clearly stated that the observed disturbances in reaching could not be explained by ataxia due to somatosensory defects. None of his patients suffered lack of position sense.

The other aspect attributed by Holmes to an impairment in visual orientation was the inability of the same patients 'to determine, or at least recognize correctly, the relative position of objects within [their] field of vision'. The patients could not describe the position of one object with respect to another, determine their respective size, or count small objects (e.g. coins) placed in front of them (some objects were not counted, others were counted twice). The fact that the patients could not appreciate the mutual spatial relations of several objects was interpreted by Holmes as reflecting in part a failure to explore the visual array by moving the eyes.

Localization of lesions responsible for these symptoms was reported by Holmes from post-mortem examination in two patients. In both cases, the entrance wound was found to have destroyed the angular and supramarginal gyri, in the posterior part of the parietal lobes on both sides. Deeper lesions were also found in the occipital lobes. Holmes discussed the possible role of the areas involved with regard to other observations in the literature (like the observation of Balint 1909), where visual disorientation had been described following similar lesions. In addition, he mentioned several observations from monkey experiments, where bilateral ablation of the 'angular gyrus' had produced misreaching for visible objects. In spite of this evidence, Holmes concluded: 'We cannot, however, assume that there exists in the regions of the angular and supramarginal gyri centres which subserve the spatial orientation of retinal impressions . . . ' (1918, p. 479). He considered that spatial orientation resulted from a 'synthesis' of visual sensations with tactile and muscular sensations, and that disorientation was due to interruption of association fibres between several brain regions.

Holmes published another case of 'disturbance of spatial orientation and visual attention' following a missile injury, and analysed it in great detail (Holmes and Horrax 1919). The patient had a homonymous visual field defect involving the inferior half of the field. Eye movements could be executed on order in all directions, but the patient had great difficulty in obtaining fixation and bringing the seen objects into central vision. Accommodation was also impaired. The most prominent sympton of the patient, according to the authors, was his inability to correctly localize in space the objects he saw in his visual field. When asked to touch any object, he pointed in a wrong direction. His errors were indiscriminately to either side of, or above or below, object position, and increased when the object was outside central vision. Figure 6.1 is a plot of the patient's pointings at targets presented at various places within his visual field. It indicates that the subject tended to reach for the target nearer his visual axis than it actually was. When the patient became able to walk, he was observed to walk into obstacles and to collide with walls. In order to avoid collisions, he progressed in small steps with his hands held in front on him like a blind person. Contrasting with the poor visual orientation, the authors noted that the patient had no problem reaching any point on his body that was touched when he had his eyes covered, and that he was able to localize sound targets in the dark.

The other aspect of visual disorientation in this patient was his inability to estimate the relative positions of two objects, especially objects located

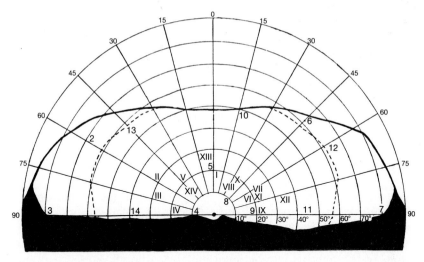

Fig. 6.1. Localization errors in a patient with a penetrating wound affecting the posterior parts of the two hemispheres. The lesion produced a loss of vision in the lower half of visual field on both sides.

The figure represents the binocular visual field. The dark area is the scotoma. A test object (a 10mm-square piece of white paper) was presented at different locations across the visual field (arabic numerals). The roman figures indicate the corresponding positions where the subject judged the test object was. As a rule the object was seen to be nearer the central part of the field than it actually was. (From Holmes and Horrax, 1919.)

at different distances from him, which he tended to see at the same distance. He also made gross errors in distinguishing the relative length or size of two objects, although his faculty to appreciate the shape of these figures was preserved. Finally, unlike the previously described patients (Holmes 1918), the present one had clear attentional and cognitive disorders. He was unable to perceive several objects at a time: only the object seen in his macular vision and on which his attention was focused could be perceived. In addition, the patient had lost topographical memory: he was unable to describe familiar places, to evoke them by imagination and to learn new routes.

There are only slight differences between the syndrome described by Balint (1909) and that described by Holmes (e.g. Holmes 1918). Predominance of visuospatial disorders in Holmes' cases and of visuomotor disorders in Balint's case is perhaps the most consistent difference. This nuance was not retained in the subsequent literature on this subject, and most of the further cases were described under the terminology of the Balint syndrome, irrespective of their dominant symptoms (see below). Kleist (1922, quoted by Brain 1941) explicitly distinguished between three components within the visual disorientation 'syndrome', which, he thought, could be observed independently from each other. These components were: defective visual localization of objects within the visual field, which Kleist attributed to visual field defects; disturbance of absolute visual localization, independent of visual field changes and leading to misdirected reaching (optic ataxia); and agnosia for visual space, which Kleist related to a loss of visual memory. This classification of symptoms in visual disorientation clearly pointed to optic ataxia as an autonomous visuomotor disorder.

6.1.3. W. R. Brain and his followers

Later authors became concerned by the fact that some of the aspects of visual disorientation appeared to be specifically related to lesions in the right hemisphere. Brain, in his paper published in 1941, was one of the first to make this point (see also Dide 1938). He reported on three patients with lesions of the right parieto-occipital junction, who all showed inattention and neglect for the left half of space, and were unable to follow familiar routes. These patients contrasted with another group of three, who showed defective visual localization (resembling optic ataxia), limited to the visual hemifield contralateral to the lesion. The lesions in these cases, however, were on the right side in one patient, and on the left side in the two others. Brain concluded that 'visual localization in external space is not a function to which dominance applies' (1941, p. 267), and that defective visual localization appears to be the result of a lesion of the upper part of the parietal lobe on either side. By contrast, what he called 'agnosia for the left half of space', he thought, was specifically related to right-sided lesions: 'No corresponding disturbance has been reported as a result of a lesion of the left hemisphere' (1941, p. 269).

Predominance of right-sided lesions in cognitive visuospatial disorders (specially the so-called 'constructional apraxia', see also Benton 1967) was confirmed by Paterson and Zangwill (1944) and Bender and Jung (1948). However, the most complete studies are those of McFie *et al.* (1950) and Hécaen *et al.* (1956) who firmly attributed the cognitive syndrome to right hemisphere lesions. The eight patients with right-sided lesions reported by McFie *et al.* (1950) had in common a left homonymous hemianopia (six of the eight cases), a neglect for the left hemispace (six of the eight cases), visuoconstructive disorders (in drawing, copying, assembling and constructing shapes, seven of the eight cases), and a loss of topographical orientation and memory (six of the eight cases). This syndrome, already described by Marie *et al.* (1922), was characterized by McFie *et al.* as a spatial 'agnosia', i.e. a coherent disorder bearing on spatial representation, specially in its relation to execution of voluntary movement; in other words, a loss of 'spatial thinking'. Although they discussed this possibility, the authors eliminated the role of eye movement disorders in producing spatial agnosia.

Hécaen *et al.* (1956) reported on 17 cases with right parietal or parieto-occipital neurosurgical ablations, of which eight were considered to present the syndrome referred to as 'apractognosia': constructional apraxia, perturbation of spatial orientation and topographical memory (in some cases only), and misjudgement of spatial co-ordinates (which Hécaen attributed to an involvement of vestibular cortical projections; see also Hécaen *et al.* 1951). The common area in the right hemisphere, defined as the area where the lesions of the eight patients overlapped, and considered as responsible for this syndrome, was centred on the posterior parietal zone (supramarginal and angular gyri and posterior part of the first temporal gyrus).

This historical review underlines several important steps in neuropsychological thinking concerning the relations of space to the brain. First, the posterior part of the parietal lobe was identified as a critical zone for mediating spatial orientation in general. Second, disorders of visual orientation were recognized to include several dissociable entities, relating either to behaviour in extrapersonal space or to cognitive representation of space. Finally, this dissociation led to assigning the right posterior parietal cortex a specific role in mediating the cognitive aspects of spatial perception and utilization. This repartition of spatial abilities between the two hemispheres is in accordance with the general view of the nature of hemispheric dominance. It is classically admitted that only abstract capacities (like language or mental images for instance), are likely to be represented unilaterally (see Bradshaw and Nettleton 1981), although operations dealing directly with the surrounding world (like visuomotor behaviour) must have an equal representation in both hemispheres. Accordingly, parietal lesion within the right hemisphere produces both spatial disorientation at the cognitive level (disorientation on a map, inability to describe spatial arrangements, constructional apraxia), and impairment in orienting behaviourally within the left hemispace. On the contrary, symmetrical lesion of the left parietal lobe produces only the behavioural disorientation within the right hemispace.

6.2. The effects of posterior parietal lesion on visually directed reaching in humans

The posterior part of the parietal lobe appears to be a critical zone for spatial localization and visuomotor behaviour. The effects of lesions restricted to this part of cortex in humans are particularly interesting to consider for understanding visuomotor mechanisms, in part because these effects are largely common to humans and monkeys. Comparison between the two species, although it will reveal marked differences, will also provide clues for correctly interpreting the effects of human lesions.

6.2.1. Optic ataxia: clinical description

In humans, lesions of the posterior parietal zone on either side may produce a striking visuomotor impairment, optic ataxia, defined as the inability to reach for objects in extrapersonal space, in the absence of gross motor, visual, or somatosensory deficits. Optic ataxia was first described in cases with bilateral involvement of the posterior parietal areas, as one of the elements of the Balint syndrome (see above). Since the initial Balint description only a few similar cases have been published. The rarity of such cases may be explained by the requirement of bilateral and more or less symmetrical lesions. Among the most clearcut cases are those of Hécaen *et al.* (1956), Hécaen and Ajuriaguerra (1954), Luria (1959), Godwin-Austen (1965), Michel *et al.* (1965), Allison *et al.* (1969), Kase *et al.* (1977), Guard *et al.* (1984), and Pierrot-Deseilligny *et al.* (1986).

Optic ataxia is now recognized as a relatively frequent deficit, which can be observed outside Balint's syndrome. Rondot (1978) has proposed the term 'visuomotor ataxia' to emphasize its specificity. Classically, patients with optic ataxia misreach for objects located within their visual field contralateral to the lesion. Misreaching can be limited to movements terminating within that part of the visual field and executed with the hand contralateral to the lesion (e.g. Castaigne *et al.* 1971), or with either hand (e.g. Garcin *et al.* 1967). Although the clinical descriptions by Castaigne's and Garcin's groups originated from patients who had lesions predominating in the posterior part of the right hemisphere, it is now generally agreed that the right hemisphere has no privilege with regard to optic ataxia.

The fact that optic ataxia often seems to be restricted to one half of the visual field (the 'visual field effect') must be interpreted with caution. A visual field defect cannot merely account for optic ataxia, for three main reasons. First, visual field is usually perimetrically normal in such patients (e.g. Damasio and Benton 1979, Garcin *et al.* 1967, Levine *et al.* 1978, Vighetto 1980). Spatial discrimination based on visual cues is also normal: patients can discriminate relative position of objects, or orientation of lines in all parts of their visual field (see Vighetto 1980). Second, as recently discovered by Levine *et al.* (1978) and Perenin *et al.* (1979; see also Ferro 1984), some patients may exhibit misreaching with the hand contralateral to the lesion in both hemifields, that is, also in the visual hemifield ipsilateral to the lesion. This 'hand effect', which seems to be more frequent in patients with lesions in the left posterior parietal area, combines with the previously described 'visual field effect', hence producing misreaching with both hands in the contralateral hemifield, and with the contralateral hand in both hemifields. Persistence of correct performance of the ipsilateral hand in the ipsilateral hemifield is a necessary but sufficient argument to disprove a purely perceptual spatial disability as a cause for optic ataxia in these patients. Finally, the third reason for not attributing optic ataxia to visual disorientation within one hemifield is that the deficit, at least under certain conditions (see below), is certainly more related to one hemispace than to one hemifield. That is, misreaching is related to spatial location of objects with respect to the body, and not merely with respect to their position on the retinal map. The routine method used for testing optic ataxia (the patient is requested to fixate the observer's nose while reaching for objects presented on either side) may have favoured the hemifield interpretation of the visuomotor defect. More complete examination reveals that optic ataxia is not limited to peripheral vision and is also present in central vision. This point is detailed in the next paragraph.

6.2.2. Constant error in pointing in optic ataxia

It is only when patients with optic ataxia are tested in a strict experimental situation that their reaching errors can be properly described and interpreted. In his thesis, Vighetto (1980) has made such an analysis in three patients (two with a left-sided lesion, one with a right-sided lesion), by using the same methodology and the same apparatus as described by Prablanc et al. (1979a; see Chapter 3). This apparatus allows manipulation of the amount of visual feedback available from the moving limb. In the 'visual-feedback' condition, in which the patients could see their moving hands during reaching, their visuomotor performances with their contralesional hand was not very different from that of normal subjects. By contrast, in the 'no visual feedback' condition, in which they could not see their hands, they became very inaccurate. Indeed, they showed the typical decrease in accuracy observed in any subject reaching for objects without being able to control his hand visually; but, in addition, the spatial distribution of their spatial errors clearly departed from that of normal subjects. The two patients with a left-sided lesion presented large constant errors, such that pointings directed at one side of midline (the side contralateral to the lesion) undershot target position, although pointings directed at targets on the opposite side overshot target position. A typical example of such a distribution is shown in Fig. 6.2 which depicts pointing errors in one patient with a vascular infarct of the left posterior parietal area (Patient Red, of Vighetto 1980, and Perenin and Vighetto 1983). This figure shows that misreaching in this patient was not merely due to inaccurate pointing, since the variable error remained within the normal range. Instead, misreaching appeared to be due to systematic deviation of pointings toward the same direction, irrespective of target position within working space. Namely, pointing was systematically deviated to the left of target positions. In the patient with a right-sided lesion, a similar but opposite deviation was

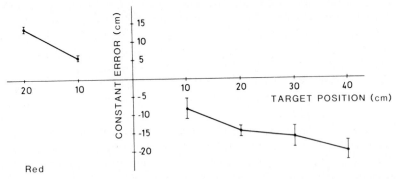

Fig. 6.2. Constant and variable errors during pointing at targets presented on either side of midline in one subject (Red) with a lesion of the posterior parietal zone of the left hemisphere. Right arm (contralateral to lesion) was used for pointing. 'No visual feedback' condition. Sign of constant error for each target position indicates systematic pointing to the left of target position. Note that the variable error is comparable to that of normal subjects in the same condition. (Replotted from Vighetto 1980).

observed. Like in the first two patients, pointings were deviated to the side of the lesion (i.e. to the right), but only within the hemispace contralateral to the lesion.

This behaviour was further documented in another patient (Biz) with a left-sided lesion. This patient was studied in a task of reaching and grasping objects placed in front of him. With his left hand (ipsilateral to the lesion), the patient performed accurate reaches in both the 'visual feedback' and the 'no visual feedback' conditions. Figure 6.3 (left) shows the points on the palmar surface of his left hand that came in contact with the target object at the end of the arm trajectories. In both conditions of visual feedback, these points appear to be located on the palmar surface of the tip of the index and third fingers, as one should expect for accurate reaching of a small object. With his hand contralateral to the lesion, however, the patient behaved in a strikingly different way whether visual feedback was available or not during the movement. In the 'visual feedback' condition, reaching was as accurate as with the normal hand, and the points of contact were restricted to the index finger (Fig. 6.3, right). In the 'no visual feedback' condition, large reaching errors appeared, the hand being systematically deviated to the left of the object (that is, to the side of the lesion), as shown by the fact that the points of contact with the object were widespread over the palmar surface of the third and fourth fingers. In addition, occasional contacts with the palm of the hand indicated a tendency of the patient to overshoot the object position in the sagittal plane (Fig. 6.3, right).

The fact that in the absence of visual feedback patients with optic ataxia

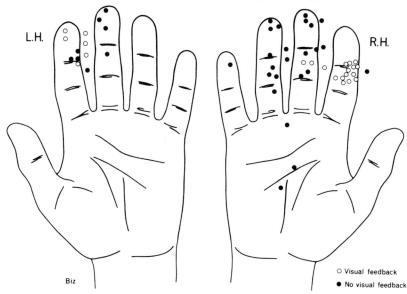

Fig. 6.3. Points of contact with object during prehension in patient Biz. L.H.: left, normal hand. R.H.: right, affected hand. Open circles: visual feedback condition. Black dots: no visual feedback condition. Points located between two fingers indicate simultaneous contact with the two fingers. Data reconstructed from film. (From Jeannerod 1986a).

misreach toward the side of the lesion when they use their hand contralateral to the lesioned hemisphere was already apparent, if not explicitly stated, in other cases described in the literature, such as those of Garcin *et al.* (1967), Tzavaras and Masure (1975), and Levine *et al.* (1978). Thus, it seems reasonable to assume that a directional error would be present in many optic ataxia patients provided they were properly examined.

Ratcliff and Davies-Jones (1972) also explicitly reported a similar finding in their extensive study of 40 subjects with chronic missile wounds involving the posterior part of one of the two hemispheres. They asked their subjects to indicate the position of stimuli presented at preselected points in their visual field, by reaching out and touching them with their index finger. Subjects with unilateral lesions (either right-or left-sided) made large errors when pointing at targets presented within the contralesional half of their visual field. Errors were more marked in the peripheral than in the central part of the field. The sign of these errors was such that the hand systematically deviated toward the side of the lesion (Fig. 6.4). Subjects with bilateral lesions tended to make localizations errors throughout the whole extent of their visual field. It is interesting to note that Ratcliff and Davies-Jones had described these errors as directed toward the patients' fixation point, rather than toward the lesion side.

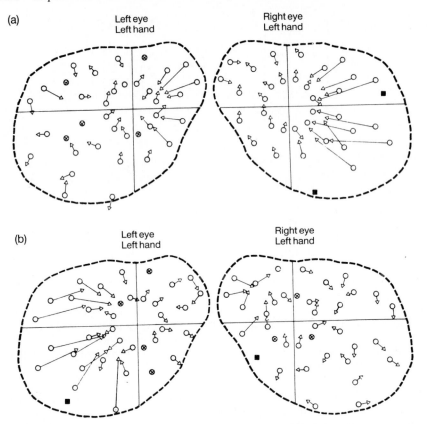

Fig. 6.4. Defective visual localization following posterior parietal lesions.

In (a) the subject had a left posterior lesion with a right hemiplegia. A 5 mm white object was projected at different locations on a spherical screen. Circles indicate positions of object. The subject pointed at these objects with his left (ipsilesional) hand. Response positions are represented by triangles. Note leftward shift of responses in the right hemifield of both eyes. Crossed circles: accurate localization. Squares: stimulus not detected.

In (b) same experiment in one subject with a right posterior lesion. In this case, the contralesional hand was used for pointing. Note rightward shift of responses in the left hemifield. (From Ratcliff and Davies-Jones 1972, with permission.)

Although this interpretation may be correct, it would not account for the fact that many optic ataxia patients, when tested in the 'no visual feedback' condition, usually misreach for objects that are presented at the body midline and therefore within the foveal visual field. In that case (for example, see Patient Biz, Fig. 6.3), their reaching movements clearly appear to be biased toward the lesion side.

One of the possible explanations for such a systematic bias is that, under conditions of unilateral parietal lesion, the patient's egocentric coordinates (those relating the spatial position of objects to the body axis and serving as a reference for directing the movements) are biased in one direction. Further arguments as to this explanation can be drawn from the observation of another type of patients with parietal lesions, those with unilateral spatial neglect (see also Chapter 4). These patients also present a strong bias toward the lesion side (usually the right side) in localization tasks such as reaching for targets (see for example cases 5 of Brain 1941, and case 3 of McFie *et al.* 1950), or bisecting a line placed in front of them. In addition, when patients with the same pathology are requested to indicate, in the dark, in which direction their body midline projects (e.g. to point at an imaginary target aligned with the middle of their chest), they also tend to err toward the side or the lesion. In the group of right hemisphere damaged patients reported by Heilman *et al.* (1983), pointings were found to be deviated by as much 8.77 cm to the right, on average (patients used their right, non paretic, hand). Perenin and Biguer (personal communication, 1986) also reported similar deviations from objective midline in a group of nine patients with right- or left-sided posterior parietal lesions (five with optic ataxia and four with unilateral neglect).

These data fit into the framework of the Kinsbourne (1970, 1987) directional hypothesis for spatial neglect. Kinsbourne assumed that neglect is not restricted to one hemispace: rather, it affects the capacity to shift attention toward the side contralateral to the lesion, whatever the absolute location in the visual field of the object toward which attention is directed. '. . . Patients with neglect', he says, 'are strikingly slow in shifting attention contralesionally, even when they initiate the shift within the intact visual half-field and direct it to a more centrally located target that is still within the same intact visual half-field'. In order to explain why neglect in humans occurs more often for the left side (as a result of right-sided lesion), Kinsbourne proposed that the 'turning tendencies' determined by each hemisphere are unequal. 'Left brain activation [according to Kinsbourne] powerfully generates rightward turning. The right brain's opposing leftward bias is weak; mainly it holds the left hemisphere's rightward bias in check. If the right hemisphere is inactivated, attention swings sharply rightward. If the left hemisphere is impaired the leftward attentional shift that results is quite mild . . ., (1987, p. 72).

It should be mentioned that the directional reaching bias is not always observed following posterior parietal lesion, and that in some cases misreaching is limited to an increase in variable error in movements executed with the contralateral arm.

6.2.3. Effects of posterior parietal lesions on movement kinematics

The effects of parietal lesions in humans are not limited to the spatial aspects of the movements. Recent studies have pointed to a much deeper motor disorganization also affecting their temporal aspects, such as movement kinematics, and co-ordination between limb segments. One must therefore consider the possibility that posterior parietal lesion alters movement programming, a function which, at first sight, would seem to be effected through the purely motor sphere.

There is information in the literature that points to such an alteration in motor programming. In optic ataxia and spatial neglect patients, the reaction time for eye and arm movements directed at targets appearing on the side contralateral to the lesion, is increased (Girotti *et al.* 1983; Perenin *et al.* 1979; Vighetto 1980). This fact has been interpreted as a partial loss of the ability to orient toward one half of space [the Heilman and Watson, (1977) 'attentional' hypothesis], or toward the side of the lesion (the Kinsbourne 'directional' hypothesis). An alternative explanation, which is developed below, proposes that the parietal lesion alters the generation, at the programming level, of spatially oriented actions in a particular direction.

Indeed, patients with optic ataxia execute visually goal-directed movements with their contralesional arm more slowly than with their ipsilesional arm. The same difference between the two arms does not appear, however, during movements directed towards non-visual targets. Patients are still able, with either arm, to make equally fast movements in actions like touching one part of their body (Vighetto 1980). This fact, together with the increase in reaction time confirms the idea that the deficit in these patients might be related specifically to the mechanisms of programming visually goal-directed movements.

In order to test this hypothesis, movement kinematics were analysed in some detail in two patients with optic ataxia following a posterior parietal lesion (Patients Biz and Tho; see Jeannerod 1986a). In both patients, neurological examination had ascertained absence of somatosensory impairment, particularly for what concerned position sense, which was normal for both hands. The technique and experimental conditions described in Chapter 2 were used. As with normal subjects, this experiment involved presentation of the objects to be reached in a sagittal plane corresponding to the body midline, at a distance of 40 cm from the body. Head movements were restrained, eye movements were free. Movements of each hand were filmed at 50 frames s^{-1}, and subsequently processed for kinematic measurements.

The results of the experiment showed striking differences between the movements of the two arms. In both patients, movements of the arm ipsilateral to the lesion were close to normal. The velocity profiles of one of these movements executed in the 'no visual feedback' condition is shown in Fig. 6.5(a) for patient Tho. Concerning Patient Biz, the velocity profiles of the ipsilesional arm trajectories in the 'no visual feedback' conditions were not entirely normal in that they involved times to peak-velocity somewhat

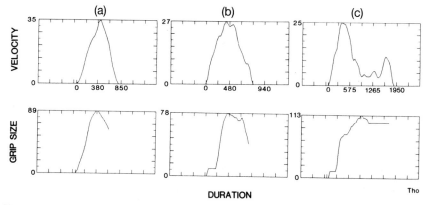

Fig. 6.5. Change in velocity of the transportion component and grip size of the manipulation component as a function of movement duration in patient Tho. (a) Normal hand, 'no visual feedback' condition. (b, c) Affected hand in the 'visual feedback' and 'no visual feedback' conditions, respectively. Maximum grip size: (a) 88.75mm; (b) 78mm; (c) 113mm. (From Jeannerod 1986a.)

Table 6.1. Patient Biz.

	Normal hand	Affected hand	
	No visual feedback	Visual feedback	No visual feedback
Duration	740 (120)	1125 (86.6)	1212 (236)
Time to velocity peak	410 (105.1)	585 (104.6)	564 (137.8)
Maximum grip size	69.7 (5.1)	70.2 (5.4)	89.2 (8.6)
Final grip size	42.0 (6.3)	50.1 (4.0)	66.4 (9.1)

Duration and time to velocity peak are in ms; maximum and final grip size (before contact) are in mm. Numbers in parenthesis are SDs.

too long with respect to total movement durations (Table 6.1). This longer time was due to stepwise increase in velocity during the acceleration phase of the movements instead of the usual sharp rise in velocity (Fig. 6.6(a)).

   In the two patients movements of the arm contralateral to the lesion differed strikingly from normal. In the 'visual feedback' condition, the general pattern of the normal transportation component was still retained, though total movement duration was increased and peak velocity was decreased with respect to the normal arm (Figs 6.5(b) and 6.6(b)). Shifting to the 'no visual feedback' condition produced two main effects. First, the patients consistently misreached the target-objects by making the systematic errors as described in the previous section. Second, total duration of their movements was further increased and the kinematics of these movements were modified. The increase in duration was not due to a change in the initial part of the movement: the velocity peak was reached within a

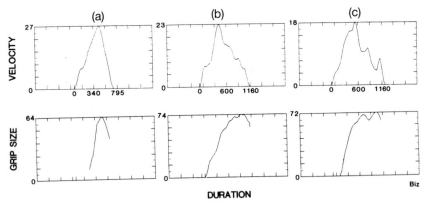

Fig. 6.6.   Change in velocity of the transportation component and grip size of the manipulation component in individual prehension movements as a function of movement duration in patient Biz. (a) Normal hand, 'no visual feedback' condition. (b, c) Affected hand, in the 'visual feedback' and 'no visual feedback' conditions, respectively. Maximum grip size: (a) 64mm; (b) 74.5mm; (c) 72mm. (From Jeannerod 1986a.)

duration close to that observed in movements of the normal hand. By contrast, the deceleration phase following the first velocity peak was dramatically increased, and several secondary peaks of smaller amplitude could be identified during deceleration (Figs 6.5(c) and 6.6(c)). These peaks were likely to correspond to an exaggeration of the normal late phase of prehension movements. Indeed, the first of these secondary peaks occurred some 600 ms after movement onset, i.e. at a time which normally corresponded to the occurrence of re-acceleration (see Chapter 2). It is interesting to note that these multiple secondary movements were observed in a situation where visual feedback from the moving hand could not be used. The implications of these results for the role of posterior parietal cortex in motor programming are discussed in another section (6.3.4).

6.2.4. Effects of posterior parietal lesion on co-ordinated hand and finger movements

Another aspect of the visuomotor deficit produced by posterior parietal lesion is impairment of finger movements. Inadequate hand and finger posturing has been reported in optic ataxia patients (Tzavaras and Masure 1975; Vighetto 1980). In order to confirm this point, film analysis was used in the same two patients for studying the manipulation components of their prehension movements. With the hand ipsilateral to the lesion, both patients performed normal grips, including in the 'no visual feedback' condition (Table 6.1 and Figs 6.5(a), 6.6(a), 6.7(a), 6.8(a)). By contrast, their hand contralateral to the lesion showed, in both conditions of visual feed-

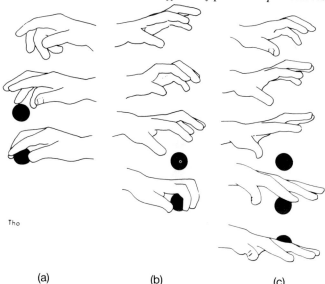

(a)	(b)	(c)

Fig. 6.7. Pattern of finger grip in patient Tho during reaching. (a) Normal hand, 'no visual feedback' condition. (b, c) Affected hand in the 'visual feedback' and 'no visual feedback' conditions, respectively. Redrawn from film. (From Jeannerod 1986a.)

back, inadequate hand shaping. In the 'visual feedback' condition, all the fingers stretched, with little evidence for grip formation until late in movement (Figs 6.5(b), 6.6(b), 6.7(b), 6.8(b)). Finger closure was incomplete and terminal grip size was too large with respect to object size (Table 6.1). In the 'no visual feedback' condition, grip formation was completely inaccurate: in patient Biz, fingers opened widely and did not close in order to accommodate the object (Table 6.1 and Figs 6.6(c) and 6.8(c)); in patient Tho, no grip formation occurred at any stage (Figs 6.5(c), 6.7(c)).

It can be argued that lack of grip formation in this type of patient is not a specific effect of parietal lesion, but could simply reflect a visuomotor 'strategy' to counteract the systematic reaching error and to optimize contact with the object. Although this might well be the case, there are arguments that point to a specific effect of the cortical lesion on grip formation. First, it can be seen from the above figures that incomplete and inadequate finger posturing in these patients was not limited to the condition where visual feedback from their hand was prevented. The deficit in grip formation was also clearly visible in the 'visual feedback' condition, i.e. in a condition where no systematic reaching error occurred. In addition, absence of finger closure during prehension is also observed in other patients with lesions outside the posterior parietal cortex, and who make no localization errors in reaching for targets (for instance patients with lesion of the motor cortex, see Chapter 2). Thus, it appears justified to conclude that the lack

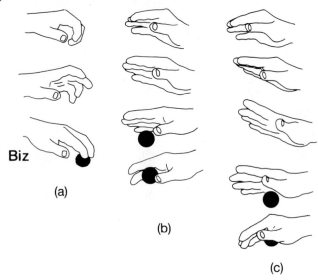

Fig. 6.8. Pattern of finger grip in patient Biz during reaching. (a) normal hand, 'no visual feedback' condition. (b, c) Affected hand in the 'visual feedback' and 'no visual feedback' conditions, respectively. Redrawn from film. (From Jeannerod 1986a.)

of grip formation in our two patients with optic ataxia when they made prehension movements with their hand contralateral to their parietal lesion, resulted from a specific alteration of visuomotor mechanisms responsible for adjusting finger posture to object shape.

6.2.5. Effects of posterior parietal lesion on eye movements

Alteration of eye movements has been mentioned many times as a possible explanation for the observed visuomotor deficits following parietal lesion. Although integrity of basic oculomotor mechanisms is usually preserved in such patients, it is the integration of eye movements into co-ordinated exploratory or reaching behaviour that is considered deficient.

In Balint's syndrome, patients have difficulty breaking visual fixation; when they succeed, their gaze errs in all directions searching for the new fixation point (Balint 1909; Holmes 1918; Michel *et al.* 1965; Pierrot-Deseilligny *et al.* 1986). Patients with unilateral spatial neglect of parietal origin may also have defective oculomotor exploration. At the early stage of the disease, the gaze is sometimes permanently deviated toward the side of the lesion, and the patients may be unable to intentionally displace their gaze beyond the midline and to explore the side opposite to the lesion. At a later stage, they progressively recover the ability to move their eyes in all directions but they may still present subtle oculomotor disorders (e.g.

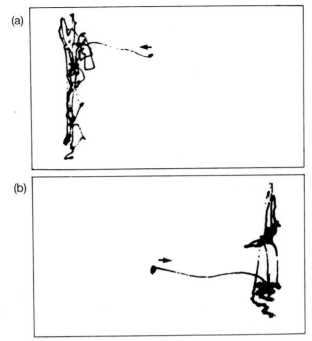

Fig. 6.9. Trajectory of gaze axis recorded during free exploration of a simple picture by one patient with a left parietal lesion (a) and one patient with a right parietal lesion (b). Exploration begins at the centre of visual field. Gaze is initially shifted in the direction marked by the arrow. Duration of exposure: 30 s. Bidimensional electro-oculography. Note consistent neglect of contralesional hemispace.

Schott *et al.* 1966). An example of such disorders is illustrated in two patients who had apparently recovered from unilateral spatial neglect. Gaze displacements were recorded two-dimensionally during examination of simple pictures. The first patient had a left-sided posterior parietal lesion with neglect of the right hemispace (an exception to Brain's 'law', see above). His eye movements were limited to exploring of the left side of the pictures, and after as long as 30 s, no eye movements were seen crossing the midline (Fig. 6.9(a)). The second patient, who had a more conventional right-sided parietal lesion and had recovered from neglect of the left hemispace, also presented asymmetrical oculomotor behaviour, by exploring only the right side of the pictures (Fig. 6.9(b)) (see Jeannerod 1985). Lack of exploration or defective scanning of visual scenes in neglect patients was also studied by De Renzi *et al.* (1970), Chédru *et al.* (1973), and Girotti *et al.* (1983).

Another impairment of eye movement in patients recovering from spatial neglect is also worth mentioning. When presented with small luminous targets appearing in the dark within the neglected hemispace, they may either fail to orient their eyes in that direction, or, after an abnormally long delay, make ocular saccades directed in the direction opposite to that of the target. This behaviour (allesthesia) illustrates the possible contribution of defective spatially oriented motor commands in producing spatial neglect (Schott *et al.* 1966).

Impaired exploratory eye movements have also been found in non-hemianopic optic ataxia patients. Vighetto (1980) and Perenin and Vighetto (1983) have observed in three such patients an increased reaction time for saccades directed to their contralesional visual field. These saccades were often fragmented in staircase patterns, similar to these observed in hemianopic patients attempting to 'explore' their scotoma (Fig. 6.10).

6.2.6. Localization of lesions responsible for optic ataxia

Lesions responsible for optic ataxia or related forms of misreaching are usually located within the upper part of the posterior parietal zone. This fact is clearly illustrated by the anatomical reconstruction of Ratcliff and Davies-Jones (1972) based on X-ray data concerning the skull alterations produced by penetrating wounds (Fig. 6.11(a)). The anatomical reconstruction of Perenin and Vighetto (1983), based on CT scan data from the lesions of five patients, also yielded the same conclusion (Fig. 6.11(b)).

More precise delimitation of the cytoarchitectonic areas involved does not seem to be possible without further improvement of imaging techniques. At present, one can only suggest that lesions producing optic ataxia would affect predominantly the superior parietal lobule (SPL), whereas lesions producing spatial neglect and spatial cognitive disorders would affect a larger area involving the inferior parietal lobule (IPL) (see Vallar and Perani 1986).

6.3. The functional significance of the posterior parietal areas

The posterior parietal cortex is one of the few parts of cerebral cortex, apart from primary areas, where a direct comparison between humans and monkeys can be attempted. The similarity of lesion effects on visuomotor behaviour in the two species calls for anatomical and functional homologies. Contribution of experimental data in the monkey thus appears essential for the understanding of human visuomotor mechanisms.

6.3.1. Anatomical considerations

Anatomical comparison of parietal lobe between monkey and human is difficult. Homologies established directly on the basis of cytoarchitectonic

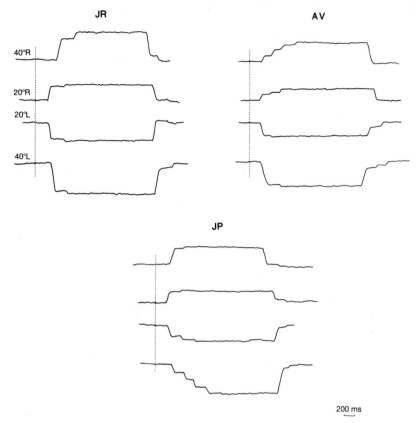

Fig. 6.10. Oculomotor responses to step visual stimuli at 20° and 40° to the right or left in three patients. Upward and downward deflections indicate movements to the right and to the left, respectively. Note the increased latencies in J.R., mostly for the 40° right target, and in J.P. for both 40° targets. Note also the staircase ocular behaviour in A.V. and J.P. when these patients have to direct the eyes at targets located in the visual field contralateral to the lesion. Vertical broken line: the moment at which targets were illuminated. Electro-oculographic technique. Lesions of these three patients have been outlined in next figure. (From Perenin and Vighetto 1983, with permission.)

maps in the two species may be misleading, specially in that part of the brain which is subject to extensive phylogenetic development.

Posterior parietal cortex involves a number of cytoarchitectonic areas defined as 'associative'. In terms of gross anatomy, this region can be subdivided into the superior and the inferior parietal lobules (SPL and IPL, respectively), separated by the intraparietal sulcus. In monkey, SPL corresponds to Brodmann's area 5, and IPL, to Brodmann's area 7 (Brodmann 1905, 1907). IPL was later split into two sub-areas, 7a and 7b (Vogt and

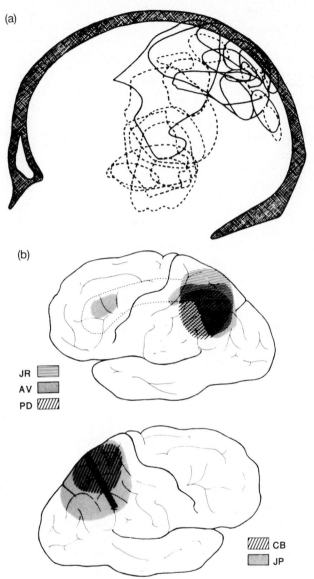

Fig. 6.11. Localization of lesions in patients with optic ataxia or related disorders.

(a) Outline of skull defects in unilateral posterior cases following penetrating wounds. Solid lines: lesions corrresponding to cases with defective localization of visual targets. Dotted lines: cases without localization problems. Note concentration of lesions in positive cases in the upper posterior parietal area. Patients without hemianopia. (From Ratcliff and Davies-Jones 1972, with permission.)

(b) Lesion reconstruction in five patients with optic ataxia, from CT scan data. The black bar on the right posterior parietal lobe represents surgical excision for removal of tumour in patient JP. (From Perenin and Vighetto 1983, with permission.)

Vogt 1919), respectively called PF and PG by von Bonin and Bailey (1947). In humans, SPL includes Brodmann's areas 5 and 7, and IPL, Brodmann's areas 39 and 40 (von Economo's areas PF and PG).

More recently, Eidelberg and Galaburda (1984) have re-examined the problem of cytoarchitectonic mapping of the parietal lobules in man. The SPL was found to be a relatively homogenous region, where three areas could identified (PE, OPE, and IPS). A fourth area (PEG) sharing the cytoarchitectonic 'style' of SPL was found at the border between SPL and IPL. IPL itself was found to be considerably more variable than SPL. It could be subdivided into five areas (PF, PFG, PG, PGH, and OPG). Eidelburg and Galaburda undertook systematic measurement of the volume of cortex corresponding to each area for comparing the extent of parietal cortex in right and left hemispheres. Area PEG (which, in terms of gross anatomy, seems to correspond to the so-called angular gyrus) was found to be larger in the right hemisphere as compared to the left. No systematic right--left asymmetry could be found for the other areas.

Thus, the problem remains to establish some correspondence between monkeys and humans. According to the view first expressed by Critchley (1953), the monkey IPL would corrrespond to the human area 7, although human areas 39 and 40 would be either absent in the monkey or present in a rudimentary form, buried within the superior temporal sulcus. This conception seems to be shared by Eidelberg and Galaburda (1984) on the basis of their cytoarchitectonic results. These authors made the important conclusion that, while human parietal lobe is distinguished in its massive expansion in size, in the elaboration of subregions and in right–left asymmetry, it also shares many of the characteristics of the monkey parietal lobe. Therefore, 'since architectonically homologous cortices . . . can be identified in both monkey and human brains, it seems likely that the posterior parietal cortices in the human brain share, at least in part, the connectional relationships found experimentally for this region in the monkey.' (Eidelberg and Galaburda 1984, p. 849).

These connections will be detailed for the monkey IPL (area 7). The main reason for this choice is that lesions of this area reproduce the human optic ataxia almost perfectly in the monkey, which suggests that the IPL connections are also shared by the critical area for optic ataxia in man. Connections of the monkey IPL are summarized in Fig. 6.12 [for a complete description, see Humphrey, (1979) and Hyvärinen (1982)]. This pattern of connectivity confirms the associative character classically attributed to this area. According to Jones and Powell (1970), area 7 would be a high-level associative area specialized in processing information within the somatosensory modality. The Jones and Powell model implies that sensory processing for each modality is elaborated within superimposed sensory--motor loops characterized by increasing degrees of associativeness. Indeed, in the somatosensory modality, primary areas (areas 1, 2, and 3)

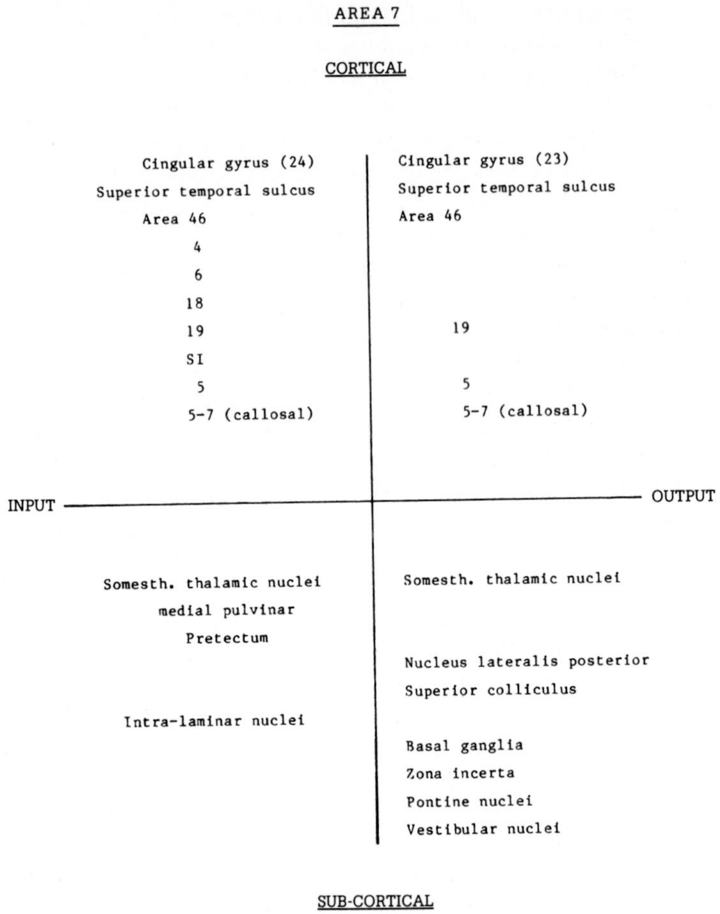

AREA 7

CORTICAL

Cingular gyrus (24)	Cingular gyrus (23)
Superior temporal sulcus	Superior temporal sulcus
Area 46	Area 46
4	
6	
18	
19	19
S I	
5	5
5-7 (callosal)	5-7 (callosal)

INPUT ─────────────────────────┼───────────────── OUTPUT

Somesth. thalamic nuclei	Somesth. thalamic nuclei
medial pulvinar	
Pretectum	
	Nucleus lateralis posterior
	Superior colliculus
Intra-laminar nuclei	
	Basal ganglia
	Zona incerta
	Pontine nuclei
	Vestibular nuclei

SUB-CORTICAL

Fig. 6.12. Diagram illustrating cortical and subcortical connections of the inferior parietal lobule in monkey. Left side: afferent connections. Right side: efferent connections.

are connected with motor and premotor cortex, while associative areas (areas 5 and 7) are connected with prefrontal cortex, superior temporal sulcus, and cingular gyrus. The same hierarchical organization would apply to visual and auditory modalities where areas 21 and 22, respectively, would play the same role as area 7 in the somatosensory modality (see Pandya and Seltzer 1982; Baleydier and Mauguière 1980). In other words, the more an area would be associative, i.e. remote from primary sensory input, the more it would be connected with the pre-motor and 'supramotor' cortex.

This model, however, does not integrate the many other connections of area 7. As can be seen from Fig. 6.12, the monkey's IPL receives a massive input from visual associative areas, as well as from other cortical associative areas (e.g. the cingular gyrus or area 46). In addition, it receives abundant input from subcortical structures such as intralaminar nuclei, medial pulvinar, somesthetic thalamic nuclei. These connections provide substrates for relatively direct somatosensory (tactile and kinaesthetic), visual and oculomotor input. At the output side, besides its already mentioned intracortical projections, IPL projects subcortically to motor pathways through the basal ganglia, the pyramidal tract, the pontine nuclei, and the vestibular complex.

The functional pattern which emerges from this description is strikingly different from the Jones and Powell model. IPL becomes a multimodal associative area contributing to the origin of a secondary, parallel, motor pathway. Thus, it can be suggested that IPL would be part of a widely distributed system including other subsystems like area 46, STS, the cingular gyrus, characterized by a supramodal status, a high degree of associativeness and a common converging output for the control of motor behaviour. Within such a system, IPL would appear more specialized for the control of spatially oriented movements. The prevalence of the visual modality in spatial organization of behaviour in primates would account for the strong projections of both striate and prestriate cortices to IPL. Although this model favours the role of area 7 as a 'spatial generator' (Jeannerod 1985), there are other conceptions as well, for instance those putting the emphasis on the role of area 7 in attention (e.g. Mesulam and Geschwind 1978; Heilman *et al.* 1987).

6.3.2. Neurophysiological aspects of posterior parietal function

The use of behaving monkeys in neurophysiological experiments has made possible the description of functional neuronal properties related to spontaneous or triggered motor activity of the animal. This type of preparation adds to the classical preparations where 'passive' neuronal properties (e.g. the receptive fields) have been described. New classes of cells are identified, which encode not only physical dimensions of the stimuli but also the behavioural and cognitive context in which stimuli are presented. In the posterior parietal areas, at least two broad classes of neurons, sensorimotor and sensory, can be distinguished.

Sensorimotor neurons are those that are selectively activated during one aspect of visuomotor behaviour, such as reaching, manipulation, visual fixation, or eye movement. Neurons related to reaching or manipulation were described by Hyvärinen and Poranen (1974) and by Mountcastle *et al.* (1975) in areas PE, PF and PG. In area PE, such cells represent 10 per cent of the total population (Mountcastle *et al.* 1975). They have no receptive field or other sensory properties in the visual or somatosensory modalities:

instead, they discharge during active reach of objects of motivational interest (a piece of food for example), or during active manipulation of these objects. Hyvärinen and Poranen (1974) considered that activation of such neurons required association of a visual stimulus and of a movement toward it. Neither presentation of the visual stimulus, nor execution of the movement alone were found sufficient conditions for firing these neurons.

Visual-fixation and eye-movement-related neurons were described by Mountcastle *et al.* (1975) and Lynch *et al.* (1977) in the monkey IPL. A large proportion of these neurons were continuously active as long as the monkey looked at an interesting target, provided it was placed within reach of the animal; firing stopped if the object was moved beyond reach (see also Hyvärinen and Poranen 1974), and could not be elicited by vigorous but uninteresting stimuli. Other neurons were related to oculomotor activity. They were activated either during tracking of continuously moving targets or in relation to saccades. The motivational value of the target was also critical. The response of these neurons was usually more pronounced for one preferential direction of the eye movement. Finally, those which were related to saccades were found to start firing before the onset of the saccade (Yin and Mountcastle 1977), hence fulfilling the criterion for 'premotor' neurons.

Posterior parietal *sensory neurons* are neurons with an identifiable receptive field. Mountcastle *et al.* (1975) found that a large proportion of the recorded neurons in area PE were activated by rotation of one or several joints. In areas PF and PG many neurons, beside those involved in visuomotor activities, respond to pure visual stimuli, such as small moving targets, revolving or expanding stimuli (see Robinson *et al.* 1978; Motter and Mountcastle 1981; Sakata *et al.* 1986).

Whether the two classes of neurons (i.e. sensorimotor and sensory) actually co-exist within the same parietal areas is a matter of debate. Experimental conditions used by the respective groups, favouring or discouraging active involvement of the animal, might explain predominance of a given class. This debate reflects the uncertainty about the precise functional significance of IPL. Hyvärinen and Poranen (1974) and Mountcastle *et al.* (1975), who were first interested by the existence of behaviourally related cells, tended to assign IPL a role in generation and initiation of neural commands for movements directed at extrapersonal space. On the other hand, other authors (e.g. Robinson *et al.* 1978) considered that properties of the visuomotor neurons (specially of the visual-fixation cells) could well be explained by 'passive' responses to visual stimuli. Accordingly they proposed that IPL was specialized in high-order sensory (visual) processing and had a predominant role in visual and visuospatial functions. In order to account for the fact that discharge of IPL neurons does increase when movements are actively produced by the animal during presentation of the stimuli, Robinson *et al.* (1978) suggested that neuron activity was 'enhanced' by its involvement in attentional mechanisms. It remains, as Lynch (1980) argued, that the fact that a given cell receives a sensory input does not preclude its participation to sensorimotor or even motor processes.

6.3.3. The effects of lesion of the inferior parietal lobule in monkey

The possibility to make surgical ablation of precisely defined cortical areas and to study their effects in performing monkeys, allows one to make relatively firm conclusions on the function of these areas. This technique has been applied with success to parietal and neighbouring areas.

(a) Dissociation between visual and spatial impairment

One problem raised by posterior parietal lesions is to distinguish between impairments affecting spatial processes involved in goal-directed actions, from impairments in visual processes, largely independent from the subject's action and involved in the perception of spatial relationships between objects. It is important to determine the possible contribution of high order perceptual deficits, particularly in the visual modality, to the disorganization of spatial orientation. As stressed earlier in this chapter, this distinction is essential for understanding the human spatial disorientation syndrome.

It has been clearly established that large posterior parietal lesions (including IPL) do not affect complex visual discriminations (e.g. Blum *et al.* 1950). According to Pohl (1973) lesions producing impairments in spatial orienting (in frontal and parietal cortex) are located differently from those (in the inferotemporal cortex) that produce impairment in non-spatial visual discrimination. In addition, monkeys with lesion of the IPL are unable to perform the task of following a bent wire track by hand (an analogue of maze tests used in humans), although they show no impairment in pattern discrimination (Petrides and Iversen 1979).

Experimental evidence for dissociating between impairments in the use of 'allocentric' and 'egocentric' spatial cues is less clear. Pohl (1973) was able to show that monkeys with a lesion of the IPL and the pre-occipital cortex were impaired in a task involving discrimination of the relative spatial location of a landmark in a visual array, but not in a task involving the same position discrimination with respect to their bodies. Animals with dorsolateral frontal lesions showed the reverse deficit, i.e. they were able to perform the 'allocentric' task, not the 'egocentric' one. More recently, Mishkin and Ungerleider (1982) and Mishkin *et al.* (1982) showed that visual modality-specific spatial abilities are subserved by the pre-occipital zone, although the IPL would subserve a supramodal spatial ability to which the visual modality (as well as other modalities) would contribute.

The effects of simultaneous ablation of the two areas combine to produce spatial disorientation (Milner *et al.* 1977; Mishkin *et al.* 1982).

It is worth mentioning at this point that clinical observation has also provided strong indications as to dissociation of the two modes of spatial processing. The classical observations reported in the historical section have revealed that patients may be unable to perform correctly oriented reaching or pointing (based on egocentric cues), although they have less difficulty in judging the relative position of objects (based on allocentric cues). The reverse is also true, i.e. patients may have a visuospatial deficit without having difficulty in reaching for objects. More recent findings have added new evidence: for instance, the typical inability of patients with posterior parietal lesions to orient in a visual maze (e.g. Benton *et al.* 1963) can be observed in the absence of other perceptual deficits. Newcombe and Russell (1969) have also shown that, according to the precise locus of the lesion in the posterior parietal zone, patients may be unable, either to recognize faces, or to resolve a maze, but that there is little or no overlap between the two deficits.

(b) Visuomotor impairment

Studies focusing on the effects of IPL lesions on visuomotor behaviour in monkey have revealed impairments that closely resemble those observed in optic ataxia patients.

The reaching deficit, first mentioned by Peele (1944), has been amply confirmed. Reaching impairment following IPL lesion in monkey is characterized by the fact that the 'hand effect' always predominates. Monkeys misreach with their arm contralateral to the lesion in either part of the visual field (Hartje and Ettlinger 1973; Faugier-Grimaud *et al.* 1978; Lamotte and Acuna 1978). Their ipsilesional arm is usually not affected, though a 'visual field effect' (involving also misreaching with the ipsilesional arm within the contralesional field) has been observed by Stein (1978). Misreaching involves a systematic bias of reaches toward the side of the lesion, a fact reported by all the above-mentioned authors. Finally, the reaching deficit is more severe in the absence of visual feedback from the limb (e.g. in darkness) than under visual guidance.

A recent study by Faugier-Grimaud *et al.* (1985) in four monkeys with a two-stage lesion of the IPL on both sides has confirmed the previous results and added new data on movement kinematics. Following the first lesion, the animals displayed a typical misreaching with the contralateral arm on both sides of the working space (Fig. 6.13). Errors were directed toward the side of the lesion. The latency of movements with the contralateral arm was increased with respect to pre-operative controls (see also Lamotte and Acuna 1978 and Stein 1978). The analysis of movement kinematics, however, revealed a striking difference with respect to human data. Following the IPL lesion in monkeys, the velocity of reaching movements executed

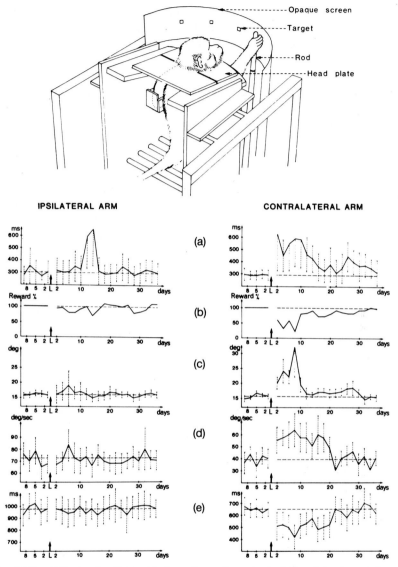

Fig. 6.13. Effects of unilateral lesion of area 7 on an aiming task in the monkey.

Drawing at the top describes the apparatus. The monkey displaces a vertical bar in front of a target, by using the arm ipsilateral or contralateral to the lesion. Accurate aiming is rewarded by food.

Results are shown for both arms on 3 days before and up to 1 month following lesion (L, arrow on time axis) in one monkey. Movements toward one target located in contralesional space are represented. (a) Latency; (b) per cent of rewarded trials (which is indicative of accuracy); (c) amplitude; (d) peak velocity; (e) duration of movements. Note effects on the contralateral arm only. (From Faugier-Grimaud *et al.* 1985, with permission.)

with the contralesional arm increased, and their duration decreased. This is the reverse of what has been described in humans following a posterior parietal lesion. In order to explain this difference one has to consider first that the available data in humans and in monkeys concern different types of movements. In humans, kinematic data were obtained with three-dimensional 'free' reaching movements, while in monkeys Faugier-Grimaud *et al.* (1985) studied one-dimensional lever displacements. The large difference in load of the moving segments between the two conditions might have affected the mode of programming and of control of the movements. It may also be that monkeys use a predominantly ballistic strategy for reaching the targets, with little participation of visual feedback and little or no correction in the terminal phase of the trajectory.

IPL ablation in the monkey also alters visually guided prehension movements. Haaxma and Kuypers (1975) first reported this fact in animals with a transcortical section involving the IPL, and a section of the corpus callosum. When placed in a situation where they had to use a precise finger grip to grasp small pellets of food, these animals were found unable to adequately shape their hand, and to achieve the grasp (see Chapter 2). The same impairment was later reported by Faugier-Grimaud *et al.* (1978) in monkeys following lesion restricted to area 7. In attempting to reach for food with their contralesional hand, these animals kept their fingers stretched and made an awkward palmar grasp instead of the normal finger grip (Fig. 6.14).

Finally, disorders of eye movements have also been reported following posterior parietal lesion in monkey. Lynch and McLaren (1979; see Lynch 1980) have shown an increased latency for saccades directed to stimuli appearing in the visual field opposite to the lesion. Bilateral lesion produced increase in latency in all directions. Latency of smooth pursuit eye movements was also increased. Increased oculomotor delays might contribute to the impaired visual search observed after posterior parietal lesion (Latto 1976). It might also represent a minor form of spatial neglect. In fact, the typical neglect syndrome, as observed in humans following right-sided posterior parietal lesions, seems to be inconsistently obtained in the monkey (Denny-Brown and Chambers 1958; Valenstein *et al.* 1982).

6.3.4. The nature of the reaching deficit produced by posterior parietal lesion

Several hypotheses can now be discussed concerning the deficit in goal-directed movements following posterior parietal lesions, whether emphasis is put on alteration of visual, somatosensory, or motor mechanisms. In this paragraph, each hypothesis is discussed, using arguments drawn from clinical observations in humans, as well as from experimental data in monkeys.

The 'visual' hypothesis relies on the important implication of monkey

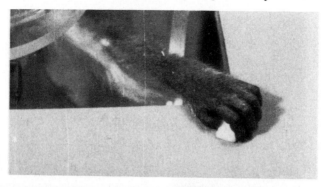

(a)

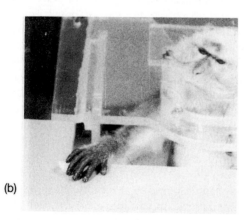

(b)

Fig. 6.14. Lack of grip formation during prehension following ablation of area 7 in monkey. (a) Normal hand ipsilateral to the lesion. (b) Affected hand contra-lateral to the lesion. Stills from film of prehension movements. (Courtesy of S. Faugier-Grimaud.)

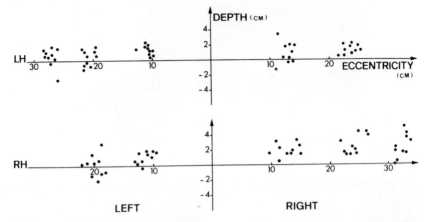

Fig. 6.15. Lack of systematic pointing bias following left parietal lesion with somato-sensory involvement in patient R.S. The figure represents a two dimensional scatter of hand pointings at visual targets in the 'no visual feedback' condition.

Upper row, left hand (LH) pointings at left and right hemispace. Lower row, right hand (RH) pointings. Note increase in variable error with the right (contralesional) hand whereas pointing bias is the same as with the normal hand. (From Jeannerod *et al.* 1984.)

IPL in visual mechanisms, as revealed by anatomical and neurophysiological data. In addition, it must be said that clinical observation in man has pointed to the existence of visuospatial deficits following lesions (particularly right-sided) in this area. Against the 'visual' hypothesis, however, is the fact that, in humans as well as in monkeys, optic ataxia or similar deficits can be observed in the absence of visuospatial problems, even when misreaching predominantly affects the contralesional visual field (the so-called 'visual-field effect'). This point will be further discussed below.

The 'somatosensory' hypothesis is probably more easily disproved than the 'visual' one. Posterior parietal lesion in humans usually does not produce tactile or proprioceptive deficits. Comparison of patients with posterior and anterior parietal lesions in fact reveals a clear double dissociation between somatosensory involvement and reaching deficits. The patient R.S. observed by Jeannerod *et al.* (1984), whose parietal lesion was limited to primary areas and to area 5, did not present any systematic reaching bias. Fig. 6.15 shows that her right (affected) arm was not significantly poorer than her left arm in pointing at targets, even in the 'no visual feedback' condition. The increase in variable error for both arms during pointing at targets within the right visual field was likely to be due to a partial hemianopia on that side.

The 'motor' or perhaps 'sensorimotor' hypothesis must be discussed in greater detail. Indeed, in the monkey, deficits of movements of the contralesional arm in all parts of the visual field stress a primarily motor impairment. This 'hand effect' is easily explained by both the anatomical connectivity of the IPL, and some of its neurophysiological characteristics (the sensorimotor neurons). The difference between monkey and human (where the 'visual-field effect' is more commonly observed than the 'hand effect'), concerning the motor involvement in posterior parietal lesion, is not as clear-cut as has been claimed. A clear 'hand effect' can be observed in some patients, particularly with left-sided lesions (Levine *et al.* 1978; Perenin *et al.* 1979; Vighetto 1980). Interestingly, this is what one would expect from the classical notion of a predominant involvement of the left hemisphere in motor impairments like limb apraxia, a disorder actually sometimes found in association with optic ataxia. In addition, the findings that in optic ataxia following both right- and left-sided lesions, the kinematics of the contralesional arm can be altered during reaching (Jeannerod 1986a), and the reaction time of movements executed with that arm is increased, are strong arguments in favour of the 'motor' hypothesis.

The level at which motor control is disorganized in optic ataxia in man or in its monkey equivalent, remains an open question. It can be tentatively suggested that the lesion in posterior parietal areas prevents access of visual information, critical for the programming of goal-directed movements, to premotor and motor structures (see Humphrey 1979). The concepts of 'hypokinesia' and 'deficit in intention' used by Heilman (see Heilman *et al.* 1987) to account for some aspects of spatial neglect might also relate to the same difficulty of initiating movements with a given arm toward a given region of extrapersonal space. This hypothesis is further discussed in the final section.

6.4. A cortical network for the control of visually guided reaching

The effects of posterior parietal lesion in human and monkey provide a basis for understanding the normal mechanisms that underly visually guided reaching. In this section, an attempt will be made at interpreting these effects as specific consequences of dysfuntioning within an identifiable cortical network. This mode of reasoning has already produced significant and testable hypotheses for deciphering complex neurological syndromes produced by cortical lesions.

The basic requirement for this approach is segmentation of visuomotor behaviour into operators, which are believed to represent actual steps in the natural function of orienting or reaching. Several of these operators have been identified throughout this book, including representation of the goal of the action, motor program, position sense, visual map, proprioceptive map, eye–head position signal, idiotropic vector, etc. Obviously those are highly speculative entities in anatomical or neurophysiological terms. They are much less so, however, in functional terms, that is, in terms of

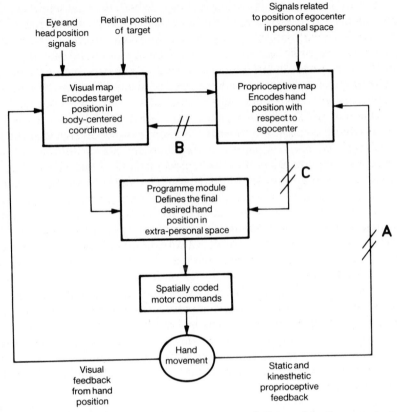

Fig. 6.16.   Flow-chart model for visuomotor control. For explanations, see text.

what functions have to exist in order to account for the observable effects
of experimental conditions or lesions. In addition, and somewhat paradoxi-
cally, connections between operators are much easier to substantiate than
the operators themselves, particularly to explain the effects of lesions. This
seems to be the case for other similar models in neuropsychology, hence
giving its heuristic value to the concept of disconnection and to the explanatory approach based on this concept (e.g. Geschwind 1965).

Some of the main operators of visuomotor behaviour and their connections have been outlined in the flow chart model of Fig. 6.16 (Jeannerod
1986c). Connections do not necessarily imply a law of succession for activation of the various operators, which may be accessed simultaneously by the
relevant signals. A feature central to the model is that the motor program
operator is permanently fed with signals from at least two others, herein
called visual map and proprioceptive map, respectively. The visual map
represents mechanisms that encode target position in space with respect to
the body. It receives signals related to retinal position of the tar-

get, and eye position in the orbit. The visual map also monitors visual signals related to the hand position in working space. In theory, the visual map would be sufficient by itself for controlling actions like reaching for a visual goal, but there are restrictions, based on the slowness and the poor specificity of visual feedback for fine control of movement (see below).

The proprioceptive map represents mechanisms that encode static and dynamic proprioceptive signals from the limb involved in the task. This encoding has to take into account other signals related to the idiotropic vector, conceived here as an internal representation of the body midline or longitudinal axis.

A second important feature of the proposed model is that visual and proprioceptive maps are interconnected. Signals related to hand position with respect to the body, and to the visually coded hand position can therefore be transferred between the two maps. This transfer of signals is of particular importance in situations where the proprioceptive map is disconnected from its normal proprioceptive input. In such situations, the only way for the motor program to be informed as to execution of the movement will be via the visual map.

Visual and proprioceptive maps jointly exert a steering influence on the motor program. Provided the two maps remain interconnected, they contain all the information needed to accurately direct the hand at the target location. This is the basis for a visually calibrated position sense, which allows feed-forward control of movements.

Some of the deficits observed in the above patients can be tentatively explained by this model. Disconnection between the proprioceptive and the visual maps (lesion B in Fig. 6.16) on one side might account for optic ataxia. The consequence of this lesion would be that, in the absence of visual feedback, no information would be available at the level of the visual map on hand position with respect to the body. In addition, perturbations of the idiotropic vector (another possible consequence of unilateral parietal lesion) would influence the action of the propriceptive map on the motor program without being checked and corrected by the visual map. This situation would produce the systematic directional bias of the hand during reaching.

Somatosensory deafferentation at the peripheral level (lesion A in Fig. 6.16) will disconnect the proprioceptive map from its normal input. As mentioned above, the consequence of this lesion is that the motor program will have to rely on the signals coming from the visual map in order to be informed as to the degree of execution of the movement. In the absence of visual feedback, the motor performance will degrade rapidly. Proprioceptive input seems to be critical, especially for actions in which several segments are involved simultaneously or in succession. Logically, this type of lesion will produce greater alteration of the most complex movements, and visual feedback will allow only an incomplete restoration.

Finally, the clinical picture of the patient R.S. (Jeannerod *et al.* 1984) suggests another interesting speculation. In this patient the cortical lesion had produced a somatosensory deafferentation syndrome similar to that observed after peripheral lesion. At variance with patients deafferented at the peripheral level, however, the R.S. nervous system received a large amount of movement-related proprioceptive signals (e.g. at the brainstem or cerebellar levels). Because these signals were ineffective in controlling the movements, one has to conclude that the proprioceptive input must reach the cortical level to influence the motor program. The most likely hypothesis for explaining the deficit of this patient would be that of a disconnection between the proprioceptive map and the motor program (lesion C in Fig. 6.16.). Persistence of a functional visual map accounted for preservation of the ability to locate targets and to make relatively accurate reaching.

These considerations stress the imitations of the feed-forward mode of control of movements. This mechanism in fact seems to require constant reafferentation by peripheral signals related to the execution and the effects of the movement. This may be because motor programs have a limited storage capacity and have only little autonomy in the time domain. When reafferentation of the motor program is no longer possible, due to pathological conditions, the system will have to shift from the optimal, feed-forward, mode of motor control to a much less efficient mode based on peripheral feedback and the overall performance will deteriorate.

References

Abbs, J.H. (1982). A speech-motor-system perspective on nervous-system-control variables. *Behav. Brain Sci.* **5**, 541–2.

—— and Gracco, V.L. (1981). Compensatory responses to low magnitude loads applied to the lower lip during speech. *J. Acoust. Soc. Am.* **70**, Suppl. 1, S78.

—— and Gracco, V.L. (1984). Control of complex motor gestures: orofacial muscle responses to load perturbations of lip during speech. *J. Neurophysiol.* **51**, 705–23.

Abend, W., Bizzi, E., and Morasso, P. (1982). Human arm trajectory formation. *Brain* **105**, 331–48.

Abrams, R.A., Kornblum, S., Meyer, D.E., and Wright, C.E. (1983). Fitts' law: optimization of initial ballistic impulses for aimed movements. *Bull. Psychon. Soc.* **22**, 335.

Adams, J.A. (1971) A closed-loop theory of motor learning. *J. Mot. Behav.* **3**, 111–49.

—— (1976). Issues for a closed-loop theory of motor learning. In: *Motor control, issues and trends* (ed. G. E. Stelmach) pp. 87–107. Academic Press, New York.

—— (1977). Feedback theory of how joint receptors regulate the timing and positioning of a limb. *Psychol. Rev.* **84**, 504–23.

Adler, F.M. (1943). Pathologic physiology of convergent strabismus. *Arch. Ophthal.* **33**, 362–77.

Alderson, G.J.K., Sully, D.J., and Sully, H.K. (1974). An operational analysis of a one-hand catching task using high-speed photography. *J. Mot. Behav.* **6**, 217–26.

Allik, J., Rauk, M., and Luuk, A. (1981). Control and sense of eye movements behind closed eyelids. *Perception* **10**, 39–51.

Allison, R.S., Hurwitz, L.J., Graham White, J., and Wilmot, T.J. (1969). A follow-up study of a patient with Balint's syndrome. *Neuropsychologia* **7**, 319–33.

Angel, R.W. (1974). Electromyography during voluntary movement. The two-burst pattern. *Electroenceph. Clin. Neurophysiol.* **36**, 493–8.

—— (1975). Electromyographic patterns during ballistic movement of normal and spastic limbs. *Brain Res.* **99**, 387–92.

—— (1977). Antagonist muscle activity during rapid arm movements: central versus proprioceptive influences. *J. Neurol. Neurosurg. Psychiat.* **40**, 683–6.

——, Alston, W., and Garland, H. (1970). Functional relations between the manual and oculomotor control signals. *Exp. Neurol.* **27**, 248–57.

Annett, J., Golby, C.W., and Kay, H. (1958). The measurement of elements in an assembly task. The information output of the human motor system. *Q. J. Exp. Psychol.* **10**, 1–11.

——, Annett, M., Hudson, P.T.W., and Turner, A. (1979). The control of movement in the preferred and non-preferred hands. *Q. J. Exp. Psychol.* **31**, 641–52.

Arbib, M.A. (1980). Interacting schemas for motor control. In: *Tutorials in motor behaviour* (ed. G. E. Stelmach and J. Requin) pp. 71–81. North-Holland, Amsterdam.

—— (1981). Perceptual structures and distributed motor control. In: *Handbook of physiology*, *Section I: The nervous system*, Vol. 2: *Motor control* (ed. V. B. Brooks) Part 2, pp. 1449–1480. Williams & Wilkins, Baltimore.

—— (1985). Schemas for the temporal organization of behavior. *Hum. Neurobiol.* **4**, 63–72.

Armand, J., Kably, B., Buisseret, P., and Moulin, G. (1986). Destruction of the proximo-distal organization of forelimb movement after neonatal sensorimotor cortex lesion in the cat. 10th Eur. Neuroscience Congr. *Neurosci. Lett.*, Suppl. 26, S84.

Ashton, J.A., Boddy, A., and Donaldson, I.M.L. (1984). Directional selectivity in the responses of units in cat primary visual cortex to passive eye movement. *Neuroscience* **13**, 653–62.

Astratyan, D.G., and Feldman, A.G. (1965). Functional tuning of the nervous system with control of movement or maintenance of a steady posture. I. Mechanographic analysis of the work of the joint on execution of a postural task. *Biophysics* **10**, 925–35.

Baker, R., Precht, W., and Llinas, R. (1972). Mossy and climbing fiber projections of extraocular muscle afferents to the cerebellum. *Brain Res.* **38**, 440–5.

Baleydier, C., and Mauguière, F. (1980). The duality of the cingulate gyrus in monkey. Neuroanatomical study and functional hypothesis. *Brain* **103**, 525–54.

Balint, R. (1909). Seelenhamung des 'Schauens', optische Ataxie, raümlische Störung des Aufmersamkeit. *Monatschr. Psychiat. Neurol.* **25**, 51–81.

Barbur, J.L. and Ruddock M.M. (1980). Spatial characteristics of movement detection mechanisms in human vision. I. Achromatic vision. *Biol. Cybern.* **37**, 77–92.

Bartz, A.E. (1966). Eye and head movements in peripheral vision: nature of compensatory eye movements. *Science* **152**, 1644.

—— (1967). Fixation errors in eye movements to peripheral stimuli. *J. Exp. Psychol.* **75**, 444–6.

Batini, C. and Buisseret, P. (1974). Sensory peripheral pathway from extrinsic eye muscles. *Arch. Ital. Biol.* **112**, 18–32.

Bauer, J.A., Wood, G.D., and Held R. (1969). A device for rapid recording of positioning responses in two dimensions. *Behav. Res. Meth. Instrum.* **1**, 157–9.

Becker, W. and Fuchs, A.F. (1969). Further properties of the human saccadic system: eye movements and correction saccades with and without visual fixation point. *Vision Res.* **9**, 1247–57.

Beggs, W.D.A. and Howarth, C.D. (1970). Movement control in man in a repetitive motor task. *Nature* **221**, 752–3.

Bender, M.B. and Jung, R. (1948). Abweichungen der Subjektiven optischen Vertikalen und Horizontalen bei Gesunden und Hirnverletzlen. *Arch. Psychiat.* **181**, 193–212.

Benton, A.L. (1967). Constructional apraxia and the minor hemisphere. *Confin. neurol. (Basel)* **29**, 1–16.

——, (1982). Spatial thinking in neurological patients: Historical aspects. In: *Spatial abilities. Developmental and physiological foundations.* (ed. M. Potegal) pp. 253–275. Academic Press, New York.

——, Elithorn, A., Fogel, M.L., and Kerr, M. (1963). A perceptual image test sensitive to brain damage. *J. Neurol. Neurosurg. Psychiat.* **26**, 540–4.

Berardelli, A., Rothwell, J.C., Day, B.L., Kachi, T., and Marsden, C.D. (1984). Duration of the first agonist EMG burst in ballistic arm movements. *Brain Res.* **304**, 183–7.

Berman, A.J. and Berman, D. (1973). Foetal deafferentation: the ontogenesis of movement in the absense of peripheral feedback. *Exp. Neurol.* **38**, 170–6.

Bernstein, N. (1967). *The coordination and regulation of movements.* Pergamon Press, Oxford.

Biguer, B., Jeannerod, M., and Prablanc, C. (1982). The coordination of eye, head and arm movements during reaching at a single visual target. *Exp. Brain Res.* **46**, 301–4.

——, Prablanc, C., and Jeannerod, M. (1984). The contribution of coordinated eye and head movements in hand pointing accuracy. *Exp. Brain Res.* **55**, 462–9.

——, Jeannerod, M., and Prablanc, C. (1985). The role of position of gaze in movement accuracy. In: *Mechanisms of attention. Attention and performance* XI, (ed. M. I. Posner and O. S. M. Marin) pp. 407–24. Erlbaum, Hillsdale, N.J.

——, Donaldson, I.M.L., Hein, A., and Jeannerod, M. (1986). La vibration des muscles de la nuque modifie la position apparente d'une cible visuelle. *C.R. Acad. Sci. Paris*, série III, **303**, 43–8.

Binet, A. and Courtier, J (1893). Sur la vitesse des mouvements graphiques. *Rev. Phil.* **35**, 664–79.

Bizzi, E., Kalil, R.E., and Tagliasco, V. (1971). Eye–head coordination in monkeys. Evidence for centrally patterned organization. *Science* **173**, 452–4.

——, —— and Morasso, P. (1972a). Two modes of active eye–head coordination in monkeys. *Brain Res.* **40**, 45–8.

——, ——, —— and Tagliasco, V. (1972b). Central programming and peripheral feedback during eye–head coordination in monkeys. *Bibl. ophthal. (Basel)* **82**, 220–32.

——, Polit, A., and Morasso, P. (1976). Mechanisms underlying achievement of final head position. *J. Neurophysiol.* **39**, 435–43.

——, Dev, P., Morasso, P., and Polit, A. (1978). Effect of load disturbances during centrally initiated movements. *J. Neurophysiol.* **41**, 542–56.

——, Chapple, W., and Hogan, N. (1982). Mechanical properties of muscles. Implications for motor control. *Trends Neurosci.* **5**, 395–8.

——, Accorneo, N., Chapple, W., and Hogan, N. (1984). Posture control and trajectory formation during arm movement. *J. Neurosci.* **4**, 2738–44.

Blum, J.S., Chow, K.L., and Pribram, K.H. (1950). A behavioural analysis of the organization of the parieto-temporo-preoccipital cortex. *J. Comp. Neurol.* **93**, 53–100.

Bock, O. and Eckmiller, R. (1986). Goal-directed arm movements in absence of visual guidance: evidence for amplitude rather than position control. *Exp. Brain Res.* **62**, 451–8.

Bonnet, C. (1982). Thresholds of motion perception. In: *Tutorials in motion perception* (ed. A. F. Wertheim, W. A. Wagenaar, and H. W. Leibowitz) pp. 41–79. Plenum, New York.

Bossom, J. (1978). Time of recovery of voluntary movement following dorsal rhizotomy. *Brain Res.* **45**, 247–50.

—— (1974). Movement without proprioception. *Brain Res.* **71**, 285–96.

—— and Ommaya, A.K. (1968). Visuomotor adaptation to prismatic transformation of the retinal image in monkeys with bilateral dorsal rhizotomy. *Brain* **91**, 161–72.

Boussaoud, D. and Joseph, J.P. (1985). Role of cat substantia nigra pars reticulata in eye and head movements. II. Effects of local pharmacological injections. *Exp. Brain Res.* **57**, 297–304.

Bower, T.G.R., Broughton, J.M., and Moore, M.K. (1970). The coordination of visual and tactual inputs in infants. *Percept. Psychophys.* **8**, 51–3.

Bowers, D. and Heilman, K.M. (1980). Pseudoneglect: effect of hemispace on a tactile line bisection task. *Neuropsychologia* **18**, 491–8.

Boyd, I.A. and Roberts, T.D.M. (1953). Proprioceptive discharges from stretch-receptors in the knee joint of the cat. *J. Physiol.* **122**, 38–58.

Bradshaw, J.L. and Nettleton, N.C. (1981). The nature of hemispheric specialization in man. *Behav. Brain Sci.* **4**, 51–92.

——, Nettleton, N.C., Nathan, G., and Wilson, L. (1983). Head and body space to left and right, front and rear. II. Visuotactual and kinesthetic studies and left-side underestimation. *Neuropsychologia* **21**, 475–86.

Brain, R. (1941). Visual disorientation with special reference to lesions of the right cerebral hemisphere. *Brain* **64**, 244–72.

Brindley, G. and Merton, P.A. (1960). The absence of position sense in the human eye. *J. Physiol.* **153**, 127–30.

Brindley G.S., Goodwin, G.M., Kulikowski, J.J., and Leighton, D. (1976). Stability of vision with paralyzed eye. *J. Physiol.* **258**, 65–6P.

Brinkman, J. and Kuypers, H.G.J.M. (1972). Split-brain monkeys: cerebral control of ipsilateral and contralateral arm, hand and finger movements. *Science* **176**, 536–9.

—— and —— (1973). Cerebral control of contralateral and ipsilateral arm, hand, and finger movements in the split-brain rhesus monkey. *Brain* **96**, 653–74.

Brodal, A. (1973). Self-observation and neuro-anatomical considerations after a stroke. *Brain* **96**, 675–94.

Brodman, K. (1905). Beiträge zur histologischen Lokalisation der Grosshirnrinde. Dritte Mitteilung: die Riedenfelder der niederen Affen. *J. Psychol. Neurol.* **4** 177–226.

——, K. (1907). Beiträge zur histologischen Lokalisation der Grosshirnrinde. Sechste Mitteilung: die Cortexgliederung des Menschen. *J. Psychol. Neurol.* **10**, 231–46.

Brouchon, M. and Paillard, J. (1966). Influence des conditions active ou passive de mobilisation d'un membre sur la précision du repérage de sa position finale. *C.R. Soc. Biol. Paris* **160**, 1281–5.

Brown, J.D., Knauft, E.B., and Rosenbaum, G. (1948). The accuracy of positioning reactions as a function of their direction and extent. *Am. J. Psychol.* **61**, 167–82.

Brown, M.C., Engberg, I.E., and Matthews, P.B.C. (1967). The relative sensitivity to vibration of muscle receptors of the cat. *J. Physiol.* **192**, 733–800.

Brown, S.H. and Cooke, J.D. (1981). Amplitude- instruction-dependent modulation of movement-related electromyogram activity in humans. *J. Physiol.* **316**, 97–107.

—— and —— (1984). Initial agonist burst duration depends on movement amplitude. *Exp. Brain Res.* **55**, 523–7.

Bruner, J.S. and Koslowski, B. (1972). Visually pre-adapted constituents of manipulatory action. *Perception* **1**, 3–14.

Buisseret, P. and Maffei, L. (1977). Extraocular proprioceptive projections to the visual cortex. *Exp. Brain Res.* **28**, 421–5.

Burgess, J.W. and Villablanca, J.R. (1986). Recovery of function after neonatal or adult hemispherectomy in cats. II. Limb bias and development from usage, locomotion and rehabilitative effects of exercise. *Behav. Brain Res.* **20**, 1–18.

Campbell, F.W. and Wurtz, R.H. (1978). Saccadic omission: why we do not see a grey-out during a saccadic eye movement. *Vision Res.* **18**, 1297–303.

Campion, J., Latto, R., and Smith, Y.M. (1983). Is blindsight an effect of scattered light, spared cortex, and near-threshold vision? *Behav. Brain Sci.* **6**, 423–86.

Campos, E.C., Chiesi, C., and Bolzani, R. (1986). Abnormal spatial localization in patients with herpes zoster ophtalmologicus. *Arch. Ophthal.* **104**, 1176–7.

Cappa, S., Sterzi, R., Vallar, G., and Bisiach, E. (1987). Remission of hemineglect and anosognosia during vestibular stimulation. *Neuropsychologia* **25**, 774–780.

Carlton, L.G. (1981). Processing visual feedback information for movement control. *J. exp. Psychol., Hum. Percept. Perf.* **7**, 1019–30.

Castaigne, P., Pertuiset, B., Rondot, P. et de Recondo, J. (1971). Ataxie optique dans les deux hémichamps visuels homonymes gauches après exérèse chirurgicale d'un anévrysme artériel de la paroi du ventricule latéral. *Rev. Neurol.* **124**, 261–8.

Chapman, E. and Wiesendanger, M. (1982). Recovery of function following unilateral lesions of the bulbar pyramid in the monkey. *Electroenceph. Clin. Neurophysiol.* **53**, 374–87.

Chédru, F., Leblanc, M., and Lhermitte, F. (1973). Visual search in normal and brain-damaged subjects. Contribution to the study of unilateral inattention. *Cortex* **9**, 94–111.

Chernikoff, R. and Taylor, F. (1952). Reaction time to kinaesthetic stimulation resulting from sudden arm displacement. *J. Exp. Psychol.* **43**, 1–8.

Clark, F.J., Horch, K.W., Bach, S.M., and Larson, G.F. (1979). Contribution of cutaneous and joint receptors to static knee-position sense in man. *J. Neurophysiol.* **42**, 877–88.

Cohen, L.A. (1961). Role of eye and neck proprioceptive mechanisms in body orientation and motor coordination. *J. Neurophysiol.* **24**, 1–11.

Conti, P. and Beaubaton, D. (1976). Utilisation des informations visuelles dans le contrôle du mouvement: étude de la précision des pointages chez l'homme. *Le Travail humain* **39**, 19–32.

Cooke, J.D. and Diggles, V.A. (1984). Rapid error correction during human arm movements. Evidence for central moitoring. *J. Mot. Behav.* **16**, 348–63.

——, Brown, S., Forget, R., and Lamarre, Y. (1985). Initial agonist burst duration changes with movement amplitude in a deafferented patient. *Exp. Brain Res.* **60**, 184–7.

Courjon, J.H., Jeannerod, M., and Prablanc, C. (1981). An attempt at correlating visuomotor-induced tilt after-effect and ocular cyclotorsion. *Perception* **10**, 519–24.

——, ——, Ossuzio, I., and Schmid, R. (1977). The role of vision in compensation of vestibulo-ocular reflex after hemilabyrinthectomy in the cat. *Exp. Brain Res.* **28**, 235–48.

Craggo, P.E., Houk, J.C. and Hasan, Z. (1976). Regulatory actions of human stretch reflex. *J. Neurophysiol.* **39**, 925–35.

Craik, K.J.W. (1947). Theory of the human operator in control systems. I. The operator as an engineering system. *Br. J. Psychol.* **38**, 56–61.

Craske, B. (1967). Adaption to prisms: change in internally registered eye position. *Br. J. Psychol.* **58**, 329–35.

Critchley, J. (1953). *The parietal lobes*. Arnold, London.

Cross, M.J. and McCloskey, D.I. (1973). Position sense following surgical removal of joints in man. *Brain Res.* **55**, 443–5.

Crossman, E.R.F.W. and Goodeve, P.J. (1983). Feedback control of hand movements and Fitts' law. Paper presented at the Meeting of Exp. Soc., Cambridge, 1963. Reprinted in *Q. J. Exp. Psychol.* (1983) **35A**, 251–78.

Damasio, A.R. and Benton, A.L. (1979). Impairment of hand movements under visual guidance. *Neurology* **29**, 170–8.

Davis, R. (1957). The human operator as a single channel information system. *Q. J. Exp. Psychol.* **9**, 119–29.

Day, B.L. and Marsden, C.D. (1982). Accurate repositioning of the human thumb against unpredictable dynamic loads is dependent upon peripheral feedback. *J. Physiol. (Lond.)* **327**, 393–407.

Denier van der Gon, J.J., and Wadman, W.J. (1977). Control of fast ballistic human arm movements. *J. Physiol.* **271**, 28–9P.

Denny-Brown, D. and Chambers, R.A. (1958). The parietal lobe and behavior. *Res. Pub. Ass. Nerv. Ment. Dis.* **36**, 35–117.

De Renzi, E. (1982). *Disorders of space exploration and cognition.* J. Wiley, New York.

——, Faglioni, P., and Scotti, G. (1970). Hemispheric contribution to exploration of space through the visual and tactual modality. *Cortex* **6**, 191–203.

Dide, M. (1938). Les désorientations temporospatiales et la prépondérance de l'hémisphère droit dans les agnoso-akinésies proporioceptives. *Encéphale* **2**, 276–95.

Di Franco, D., Muir, D.W., and Dodwell, P.C. (1978). Reaching in very young infants. *Perception* **7**, 385–92.

Di Stefano, M., Morelli, M., Marzi, C.A., and Berlucchi, G. (1980). Hemispheric control of unilateral and bilateral movement of proximal and distal parts of the arm as inferred from simple reaction time to lateralized light stimuli in man. *Exp. Brain Res.* **38**, 197–204.

Donaldson, I.M.L. and Long, A.C. (1980). Interaction between extraocular proprioceptive and visual signals in the superior colliculus of the cat. *J. Physiol.* **298**, 85–110.

Duchenne de Boulogne, G.B.A. (1855). *De l'électrisation localisée, et son application à la pathologie et à la thérapeutique.* Baillère, Paris.

Eden, A.E. and Correia, M.J. (1982). An autoradiographic and HRP study of vestibulocollic pathways in the pigeon. *J. Comp. Neurol.* **211**, 432–40.

Eidelberg, D. and Galaburda, A.M. (1984). Inferior parietal lobule. Divergent architectonic asymmetries in the human brain. *Arch. Neurol.* **41**, 843–52.

Eklund, G. (1972). Position sense, and state of contraction. The effects of vibration. *J. Neurol. Neurosurg. Psychiat.* **35**, 606–11.

Elliott, D. and Allard, F. (1985). The utilisation of visual feedback information during rapid pointing movements. *Q. J. Exp. Psychol.* **37A**, 407–25.

Evarts, E.V. (1968). Relation of pyramidal tract activity to force exerted during voluntary movement. *J. Neurophysiol.* **31**, 14–27.

—— and Vaughn, W.J. (1978). Intended arm movements in response to externally produced arm displacements in man. In: *Cerebral motor control in man: long loop mechanisms* (ed. J. E. Desmedt) pp. 178–192. *Progress in Clinical Neurophysiology* Karger, Basel.

Faugier-Grimaud, S., Frenois, C., and Stein, D.G. (1978). Effects of posterior parietal lesions on visually guided behavior in monkeys. *Neuropsychologia* **16**, 151–68.

——, ——, and Peronnet, F. (1985). Effects of posterior poarietal lesions on visually guided movements in monkeys. *Exp. Brain Res.* **59**, 125–38.

Feinberg, T.E., Pasik, T., and Pasik, P. (1978). Extrageniculostriate vision in the monkey. VI. Visually guided accurate reaching behavior. *Brain Res.* **152**, 422–8.

Feldman, A.G. (1966). Functional tuning of the nervous system during control of movement or maintenance of a steady posture. II. Controllable parameters of the muscle. *Biophysics* **11**, 565–78.

—— (1981). The composition of central programs subserving horizontal eye movements in man. *Biol. Cybern.* **42**, 107–16.

Feldman, A.G. and Latash, M.L. (1982a). Interaction of afferent and efferent signals underlying joint position sense. Empirical and theoretical approaches. *J. Mot. Behav.* **14**, 174–93.

—— and —— (1982b). Afferent and efferent components of joint position sense. Interpretation of kinaesthetic illusion. *Biol. Cybern.* **42**, 205–14.

Felix, D. and Wiesendanger, M. (1971). Pyramidal and nonpyramidal motor cortical effects on distal forelimb muscles of monkeys. *Exp. Brain Res.* **12**, 81–91.

Ferro, J.M. (1984). Transient inaccuracy in reaching caused by a posterior parietal lobe lesion. *J. Neurol. Neurosurg. Psychiat.* **47**, 1016–19.

Fessard, A. (1926–7). *Le mouvement volontaire (d'après K. Wacholder).* Cours et Conférences de la Faculté de Médecine et des Hôpitaux de Paris. Chahine, Paris.

Festinger, M.L. and Canon, L.K. (1965). Information about spatial location based on knowledge about efference. *Psychol. Rev.* **72**, 373–84.

Fiorentini, A. and Ercoles, A.M. (1966). Visual direction of a point source in the dark. *Atti. Fond. B. Ronchi.* **23**, 405–28.

——, Berardi, N., and Maffei, L. (1982). Role of extraocular proprioception in the orienting behavior of cats. *Exp. Brain Res.* **48**, 113–20.

Fischer, B. and Ramsperger, E. (1984). Human express saccades. Extremely short reaction times of goal-directed eye movements. *Exp. Brain Res.* **57**, 191–5.

—— and Rogal, L. (1986). Eye-hand coordination in man. A reaction time study. *Biol. Cybern.* **55**, 253–61.

Fisk, J.D. and Goodale, M.A. (1985). The organization of eye and limb movements during unrestricted reaching to targets in contralaterral and ipsilateral visual space. *Exp. Brain Res.* **60**, 159–78.

Fitts, P.M. (1954). The information capacity of the human motor system in controlling the amplitude of movement. *J. Exp. Psychol.* **47**, 381–91.

Flandrin, J.M. and Jeannerod, M. (1981). Effects of unilateral superior colliculus ablation on oculomotor and vestibulo-ocular responses in the cat. *Exp. Brain Res.* **42**, 73–80.

Flash, T. and Hogan, N. (1985). The coordination of arm movements: an experimentally conformed mathematical model. *J. Neurosci.* **5**, 1688–703.

Flowers, K. (1975). Handedness and controlled movement. *Br. J. Psychol.* **66**, 39–52.

Folkins, J. and Abbs, J.H. (1975). Lip and jaw motor control during speech: responses to resistive loading of the jaw. *J. Speech Hearing Res.* **18**, 207–20.

Forget, R. and Lamarre, Y. (1981). Etude chez l'homme de l'activité des muscles agonistes et antagonistes lors d'un mouvement rapide de flexion de l'avant-bras. *J. Physiol. (Paris)* **77**, 34A.

—— and —— (1982). Contribution of peripheral feedback to the control of a rapid flexion movement of the forearm in man. *Soc. Neurosci. Abstr.* **8**, 732.

Fraser, C. and Wing, A. (1981). A case study of reaching by a user of a manually-operated artificial hand. *Prosthetics and Orthotics Int.* **5**, 151–6.

Freund, H.J. and Büdingen, H.J. (1978). The relationship between speed and amplitude of the fasted voluntary contractions of human arm muscles. *Exp. Brain Res.* **31**, 1–12.

Fuchs, A. and Kornhuber, H.H. (1969). Extraocular muscle afferents to the cerebellum of the cat. *J. Physiol.* **200**, 713–23.

Fuchs, A.F. and Luschei, E.S. (1970). Firing patterns of abducens neurons of alert monkeys in relationship to horizontal eye movements. *J. Neurophysiol.* **33**, 382–92.

Gachoud, J.P., Mounoud, P., Hauert, C.A., and Viviani, P. (1983). Motor strategies in lifting movements. A comparison of adult and child performance. *J. Mot. Behav.* **15**, 202–16.

Gandevia, S.C. (1982). The perception of motor commands or effort during muscular paralysis. *Brain* **105**, 151–9.

—— (1985). Illusory movement produced by electrical stimulation of low-threshold muscle afferents from the hand. *Brain* **108**, 965–81.

—— and McCloskey, D.I. (1976). Joint sense, muscle sense and their combination as position sense, measured at the distal interphalangeal joint of the middle finger. *J. Physiol.* **260**, 387–407.

—— and —— (1977a). Effects of related sensory inputs on motor performances in man studied through changes in perceived heaviness. *J. Physiol.* **272**, 653–72.

—— and —— (1977b). Changes in motor commands, as shown by changes in perceived heaviness, during partial curarization and peripheral anaesthesia in man. *J. Physiol.* **272**, 673–89.

Garcin, R., Rondot, P., and De Recondo, J. (1967). Ataxie optique localisée aux deux hémichamps visuels homonymes gauches. *Rev. Neurol.* **116**, 707–14.

Garland, H., Angel, R.W., and Moore, W.E. (1972). Activity of triceps brachii during voluntary elbow extension. Effect of lidocaïne blockade of elbow flexors. *Exp. Neurol.* **37**, 231–5.

Gauthier, G.M. and Hofferer, J.M. (1976). Eye tracking of self-moved targets in the absence of vision. *Exp. Brain Res.* **26**, 121–39.

Gazzaniga, M.S., Bogen, J.E., and Sperry, R.W. (1967). Dyspraxia following division of the cerebral commissures. *Arch. Neurol.* **16**, 606–12.

Gelfan, S. and Carter, S. (1967). Muscle sense in man. *Exp. Neurol.* **18**, 469–73.

Gentner, D.R. (1985). Skilled motor performance at variable rates. A composite view of motor control. CHIP report 124, University of California, San Diego.

Geschwind, N. (1965). Disconnexion syndromes in animals and man. *Brain* **88**, 237–94, 585–644.

Gesell, A. (1938). The tonic neck reflex in the human infant. *J. Pediat.* **13**, 455–464.

Ghez, C. (1979). Contributions of central programs to rapid limb movement in the cat. In: *Integration in the nervous system* (ed. H. Asanuma and V. J. Wilson) pp. 305–320. Igaku-Shoin, Tokyo.

—— and Martin, J.H. (1982). The control of rapid limb movement in cat. III. Agonist-antagonist coupling. *Exp. Brain Res.* **45**, 115–25.

Gielen, C.C.A.M., Van den Heuvel, P.J.M., and Van Gisbergen, J.A.M. (1984). Coordination of fast eye and arm movements in a tracking task. *Exp. Brain Res.* **56**, 154–61.

Girotti, F., Casazza, M., Musicco, M., and Avanzini, G. (1983). Oculomotor disorders in cortical lesions in man: the role of unilateral neglect. *Neuropsychologia* **21**, 543–54.

Godwin-Austen, R.B. (1965). A case of visual disorientation. *J. Neurol. Neurosurg. Psychiat.* **28**, 453–8.

Goldscheider, A. (1898). *Gesammelte Abhandlung*. Vol. II. Physiologie des Mus-
kelsinnes, Leipzig.

Goodwin, G.M., McCloskey, D.I., and Matthews, P.B.C. (1971). A systematic
distortion of position sense produced by muscle vibration. *J. Physiol.* **218**.

——, ——, and —— (1972). The contribution of muscle afferents to kinaesthesia
shown by vibration induced illusions of movement and by the effects of paralys-
ing joint afferents. *Brain* **95**, 705–48.

Gordon, J. and Ghez, C. (1984). EMG patterns in agonist muscles during isometric
contraction in man: relations to reponse dynamics. *Exp. Brain Res.* **55**, 167–71.

Goutières, F., Challamel, M.J., Aicardi, J., and Gilly, R. (1972). Les hémiplégies
congénitales. Sémiologie, étiologie et pronostic. *Arch. franç. Pédiat.* **29**, 839–51.

Granit, R. (1972). Constant errors in the execution and appreciation of movement.
Brain **95**, 649–60.

Grenier, A. (1981). La 'motricité libérée' par fixation manuelle de la nuque au
cours des premières semaines de la vie. *Arch. franç. Pédiat.* **38**, 557–61.

Gresty, M.A. (1974). Coordination of head and eye movements to fixate continuous and intermittent targets. *Vision Res.* **14**, 395–403.

Grigg, P., Finerman, G.A., and Riley, L.H. (1973). Joint position sense after total
hip replacement. *J. Bone Jt. Surg.* **55**, 1016–25.

Grillner, S. (1985). Neurobiological bases of rhythmic motor acts in vertebrates.
Science **228**, 143–9.

Grossberg, S. and Kuperstein, M. (1986). Neural dynamics of adaptive sensory-
motor control. *Advances in Psychology*, Vol. 30. North-Holland, Amsterdam.

Grüsser, O.J. (1984). J.E. Purkinje's contributions to the physiology of the visual,
the vestibular and the oculomotor systems. *Human Neurobiol.* **3**, 129–44.

Guard, O., Perenin, M.T., Vighetto, A., Giroud, M., Tommasi, M. and Dumas,
R. Syndrome pariétal bilatéral proche d'un syndrome de Balint. *Rev. Neurol.*
(Paris) **140**, 358–67.

Guitton, D., Douglas, R.M. and Volle, M. (1984). Eye-head coordination in the
cat. *J. Neurophysiol.* **52**, 1030–50.

Guthrie, B.L., Porter, J.D., and Sparks, D.L. (1983). Corollary discharge provides
accurate eye position information to the oculomotor systems. *Science* **221**,
1193–5.

Haaxma, H. and Kuypers, H.G.J.M. (1975). Intrahemispheric cortical connections
and visual guidance of hand and finger movements in the rhesus monkey. *Brain*
98, 239–60.

Hallett, M. and Marsden, C.D. (1979). Ballistic flexion movements of the human
thumb. *J. Physiol. (Lond.)* **294**, 33–50.

——, Shahani, B.T., and Young, R.R. (1975). EMG analysis of stereotyped voluntary movements in man. *J. Neurol. Neurosurg. Psychiat.* **38**, 1154–62.

Hallett, P.E. and Lightstone, A.D. (1976). Saccadic eye movement towards stimuli
triggered by prior saccades. *Vision Res.* **16**, 99–106.

Halverson, H.M. (1931). An experimental study of prehension in infants by means
of systematic cinema records. *Genet. Psychol. Monogr.* **10**, 110–286.

Hammond, P.H. (1960). An experimental study of servo-action in human muscular
control. In: *Proc. 3rd Int. Conf. Electr.*, pp. 190–9. Institution of Medical Engineers, London.

Harrris, C.S. (1974). Beware of the straight ahead shift. A non-perceptual change
in experiments on adaptation to displaced vision. *Perception* **3**, 461–76.

Harris, L.R. (1986). The superior colliculus and movements of the head and eyes in cats. *J. Physiol.* **300**, 367–91.

——, Blakemore, C., and Donaghy, M. (1980). Integration of visual and auditory space in the mammalian superior colliculus. *Nature* **288**, 56–9.

Hartje, W. and Ettlinger, G. (1973). Reaching in light and dark after unilateral posterior parietal ablations in the monkey. *Cortex* **9**, 346–54.

Hay, L. and Brouchon, M. (1972). Analyse de la réorganisation des coordinations visuo-motrices chez l'homme. *Ann. Psychol. (Paris)* **72**, 25–38.

Head, H. (1920). *Studies in neurology*. Hodder & Stoughton, London.

Hécaen, H. (1984). *Les gauchers* (2nd edn). Presses Universitaires de France, Paris.

—— and de Ajuriaguerra, J. (1948). Etude des troubles toniques, moteurs et végé-tatifs et de leur récupération après ablation limitée du cortex moteur et prémo-teur. Congrès des Médecins Alinénistes et Neurologistes, pp. 269–74.

—— and —— (1954). Balint's syndrome (psychic paralysis of gaze fixation) and its minor forms. *Brain* **77**, 373–400.

——, ——, and Massonnet, J. (1951). Les troubles visuo-constructifs par lésion parieto-occipitale droite. Rôle des perturbations vestibulaires. *Encéphale* **40**, 122–79.

——, Penfield, W., Bertrand, C., and Malmo, R. (1956). The syndrome of aprac-tognosia due to lesions of the minor cerebral hemisphere. *Arch. Neurol. Psy-chiat.* 75, 400–34.

——, Perenin, M.T., and Jeannerod, M. (1984). The effects of cortical lesions in children; language and visual functions. In: *Early brain damage: Research orien-tations and clinical observations*, (ed. C. R. Almli and S. Finger) Vol. 1, pp. 277–298. Academic Press, New York.

Heilman, K.M. and Watson, R.T. (1977). The neglect syndrome. A unilateral defect of the orienting response. In: *Lateralisation in the nervous system* (ed. S. Harnad, R. W. Doty, L. Goldstein, J. Jaynes, and G. Krauthamer) pp. 285–302. Academic Press, New York.

——, Bowers, D., and Watson, R.T. (1983). Performance on hemispatial pointing task by patients with neglect syndrome. *Neurology* **33**, 661–4.

——, ——, Valenstein, E., and Watson, R.T. (1987). Hemispace and hemispatial neglect. In: *Neurophysiological and neuropsychological aspects of spatial neglect* (ed. M. Jeannerod) pp. 115–150. North-Holland, Amsterdam.

Hein, A. and Diamond, R. (1972). Locomotory space as a prerequisite for acquir-ing visually guided reaching in kittens. *J. Comp. Physiol. Psychol.* **81**, 394–8.

—— and —— (1983). Contribution of eye movement to the representation of space. In: *Spatially oriented behavior*, (ed. A. Hein and M. Jeannerod), pp. 119–133. Springer-Verlag, New York.

——, Vital-Durand, F., Salinger, W., and Diamond, R. (1979). Eye movement initiate visual-motor development in the cat. *Science* **204**, 1221–2.

Held, R. (1961). Exposure-history as a factor in maintaining stability of perception and coordination. *J. Nerv. Ment. Dis.* **132**, 26–32.

—— (1965). Plasticity in sensorimotor systems. *Scient. Am.* **213**, 84–94.

—— and Freedman, S. (1963). Plasticity in human sensorimotor control. *Science* **142**, 455–62.

—— and Gottlieb, N. (1958). Technique for studying adaptation to disarranged hand eye coordination. *Percept. Mot. Skills* **8**, 83–6.

Henderson, W.R. and Smyth, G.E. (1948). Phantom limbs. *J. Neurol. Neurosurg. Psychiat.* **11**, 88–112.

Henn, V. The history of cybernetiks in the 19th century. In: *Pattern recognition in biological and technical systems*, pp. 1–7. Springer-Verlag, Berlin.

Henson, D.B. (1978). Corrective saccades: effects of altering visual feedback. *Vision Res.* **18**, 63–7.

Herman, R., Herman, R., and Maulucci, R. (1981). Visually-triggered eye-arm movements in man. *Exp. Brain Res.* **42**, 392–98.

Higgins, J.R., and Angel, R.W. (1970). Correction of tracking errors without sensory feedback. *J. Exp. Psychol.* **84**, 412–16.

Hill, A.L. (1972). Direction constancy. *Percept. Psychophys.* **11**, 175–8.

Hollerbach, J.M. and Flash, T. (1982). Dynamic interactions between limb segments during planar arm movements. *Biol. Cybern.* **44**, 67–77.

Holmes, G. (1918). Disturbances of visual orientation. *Br. J. Ophthal.* **2**, 449–506.

—— and Horrax, G. (1919). Disturbances of spatial orientation and visual attention with loss of stereoscopic vision. *Arch. neurol. Psychiat.* **1**, 385–407.

Honda, H. (1984). Eye position signals in successive saccades. *Percept. Psychophys.* **36**, 15–20.

Howard, I.P. (1982). *Human visual orientation*. Wiley, Toronto.

Hulliger, M., Nordh, E., Thelin, A.E., and Vallbo, A.B. (1979). The responses of afferent fibres from the glabrous skin of the hand during voluntary finger movements in man. *J. Physiol.* **291**, 233–49.

Humphrey, D.R. (1979) On the cortical control of visually directed reaching. Contributions by nonprecentral areas. In: *Posture and movement* (ed. R. E. Talbot and D. R. Humphrey) pp. 51–112. Raven Press, New York.

Humphrey, N.K. and Weiskrantz, L. (1969). Vision in monkeys after removal of the striate cortex. *Nature* **215**, 595–7.

Humphrey, T. (1969). Postnatal repetitions of human prenatal activity sequences with some suggestions of their neuroanatomical basis. In: *Brain and early behavior* (ed. R. J. Robinson), Academic Press, London.

Hutton, J.T., and Palet, J. (1986). Lateral saccadic latencies and handedness. *Neuropsychologia* **24**, 449–51.

Hyvärinen, J. (1982). *The parietal cortex of monkey and man*. Springer-Verlag, Berlin.

—— and Poranen, A. (1974). Function of the parietal associative area 7 as revealed from cellular discharges in alert monkeys. Brain **97**, 673–92.

Jackson, J.H. and Paton, L. (1909). On some abnormalities of ocular movements. *Lancet* (27 March) 900–5.

Jagacinski, R.J., Repperger, D.W., Moran, M.S., Ward, S.L., and Glass, B. (1980). Fitts' law and the microstructure of rapid discrete movements. *J. Exp. Psychol., Human Percept. Perform.* **6**, 309–20.

James, W. (1890). *The principles of psychology*. Macmillan, London.

Jay, M.F. and Sparks, D.L. (1984). Auditory receptive fields in primate superior colliculus shift with changes in eye position. *Nature* **309**, 345–7.

Jeannerod, M. (1981a). Specialized channels for cognitive responses. *Cognition* **10**, 135–137.

——, (1981b). Intersegmental coordination during reaching at natural visual objects. In: *Attention and performance* (ed. J. Long and A. Baddeley) pp. 153–68. Erlbaum, Hillsdale, N.J.

—— (1983). *Le cerveau-machine. Physiologie de la volonté*. Fayard, Paris. English transl.: *The brain machine. The development of neurophysiological thought*, Harvard University Press, Cambridge, Mass., 1985.

—— (1984). The timing of natural prehension movements. *J. Mot. Behav.* **16**, 235–54.

—— (1985). The posterior parietal area as a spatial generator. In: *Brain mechanisms and spatial vision* (ed. D. Ingle, M. Jeannerod, and D. N. Lee) pp. 279–98. M. Nijhoff, Dordrecht.

—— (1986a). The formation of finger grip during prehension. A cortically mediated visuomotor pattern. *Behav. Brain Res.* **19**, 99–116.

—— (1986b). Are corrections in accurate arm movements corrective? In: *Progress in Brain Research* (ed. H. J. Freund, U. Büttner, B. Cohen, and J. Noth) Vol. 64, pp. 353–360. Elsevier, Amsterdam.

—— (1986c). Mechanisms of visuomotor co-ordination: A study in normal and brain-damaged subjects. *Neuropsychologia.* **24**, 41–78.

—— and Biguer, B. (1982). Visuomotor mechanisms in reaching within extrapersonal space. In: *Advances in the analysis of visual behaviour* (ed. D. Ingle, M. Goodale, and R. Mansfield) pp. 387–409. MIT Press, Boston.

—— and Prablanc, C. (1983). The visual control of reaching movements. In: *Motor control mechanisms in man* (ed. J. E. Desmedt) pp. 13–29. Raven Press, New York.

——, Gerin, P., and Mouret, J. (1965). Influence de l'obscurité et de l'occlusion des paupières sur le contrôle des mouvements oculaires. *Ann. Psychol. (Paris)* **65**, 309–24.

——, Kennedy, H., and Magnin, M. (1979). Corollary discharge: its possible implications in visual and oculomotor interactions. *Neuropsychologia* **17**, 241–58.

——, Michel, F., and Prablanc, C. (1984). The control of hand movements in a case of hemianaesthesia following a parietal lesion. *Brain* **107**, 899–920.

Jones, B. (1974). Role of central monitoring of efference in short-term memory for movements. *J. Exp. Psychol.* **102**, 37–43.

Jones, E.G. (1972). The development of the 'muscular sense' concept during the nineteenth century and the work of H. Charlton Bastian. *J. Hist. Med. All. Sci.* **27**, 298–311.

—— and Powell, T.P.S. (1970). An anatomical study of converging sensory pathways within the cerebral cortex of the monkey. *Brain* **93**, 793–820.

Kalil, R.E and Freedman, S.J. (1966). Persistence of ocular rotation following compensation for displaced vision. *Percept. Mot. Skills* **22**, 135–9.

Kase, C.S., Troncoso, J.F., Court, J.E., Tapia, J.F., and Mohr, J.P. (1977). Global spatial disorienttion. *J. Neurol. Sci.* **34**, 267–78.

Keele, S.W. (1968). Movement control in skilled motor performance. *Psychol. Bull.* **70**, 387–404.

—— (1981). Behavioural analysis of movement. In: *Handbook of physiology, Section I: The nervous system*, Vol. II: *Motor control* (ed. V. B. Brooks) Part 2, pp. 1391–414. Williams & Wilkins, Baltimore.

—— and Posner, M.I. (1968). Processing of visual feedback in rapid movements. *J. Exp. Psychol.* **77**, 155–8.

Keller, E.L. and Robinson, D.A. (1971). Absence of stretch reflex in extra-ocular muscles of the monkey. *J. Neurophysiol.* **34**, 908–19.

Kelso, J.A.S. and Holt, K.G. (1980). Exploring a vibratory systems analysis of human movement production. *J. Neurophysiol.* **43**, 1183–96.

—— and Tuller, B. (1984). A dynamical basis for action systems. In: *Handbook of cognitive neuroscience* (ed. M.S. Gazzaniga) pp. 321–56. Plenum Press, New York.

——, Southard, D.L., and Goodman, D. (1979). On the coordination of two-handed movements. *J. Exp. Psychol., Hum. Percept. Perform.* **5**, 229–38.

——, Holt, K.G., and Flatt, A.E. (1980a). The role of proprioceptive in the perception and control of human movement: toward a theoretical reassessment. *Percept. Psychophys.* **28**, 45–52.

——, Holt, K.G., Kugler, P.N., and Turvey, M.T. (1980b). On the concept of coordinative structures as dissipative structures. II. Empirical lines of convergence. In: *Turorials in motor behavior* (ed. G. E. Stelmach and J. Requin) pp. 49–70. North-Holland, Amsterdam.

——, Tuller, B., and Harris, K.S. (1981). A 'dynamic pattern' perspective on the control and coordination of movement. In: *The production of speech* (ed. P. MacNeilage). Springer-Verlag, New York.

——, Putnam, C.A., and Goodman, D. (1983). On the space-time structure of human interlimb coordination. *Q. J. Exp. Psychol.* **35A**, 347–75.

——, Tuller, B., Vatikiotis-Bateson, E., and Fowler, C.A. (1984). Functionally specific articulatory cooperation following jaw perturbations during speech. Evidence for coordinative structures. *J. Exp. Psychol., Hum. Percept. Perform.* **10**, 811–32.

Kennard, M.A.,and Ectors, L. (1938). Forced circling in monkeys following lesions of the frontal lobes. *J. Neurophysiol.* **1**, 45–54.

Kent, R.D., Carney, P.J., and Severeid, L.R. (1974). Velar movement and timing: evaluation of a model for binary control. *J. Speech Hearing Res.* **17**, 470–88.

Kinsbourne, M. (1970). A model for the mechanisms of unilateral neglect of space. *Trans. Am. Neurol. Assoc.* **95**, 143–6.

—— (1987). Mechanisms of unilateral neglect. In: *Neurophysiological and neuropsychological aspects of spatial neglect* (ed. M. Jeannerod) pp. 69–86. North-Holland, Amsterdam.

Klapp, S.T. (1975). Feedback versus motor programming in the control of aimed movements. *J. Exp. Psychol., Hum. Percept. Perform.* **104**, 147–53.

Klein, B.G., Deich, J.D., and Zeigler, H.P. (1985). Grasping in the pigeon (Columbia Livia): final common path mechanisms. *Behav. Brain Res.* **18**, 201–13.

Knapp, H.D., Taub, E., and Berman, A.J. (1963). Movements in monkeys with deafferented forelimbs. *Exp. Neurol.* **7**, 305–15.

Koerner, F.H. (1975a). Non-visual control of human saccadic eye movements. In: *Basic mechanisms of ocular motility and their clinical implications* (ed. G. Lennerstrand and P. Bach-y-Rita) pp. 565–9. Pergamon Press, Oxford.

—— (1975b). Untersuchungen über die nichtvisuelle Kontrolle von Augenbewegungen. *Adv. Ophthal.* **31**, 100–58.

Kornmüller, A.E. (1931). Eine Experimentelle Anästhesie der aüsseren Augenmuskeln am Meschen und ihre Auswirkungen. *J. Psychol. Neurol.* **41**, 351–66.

Kugler, P.N., Kelso, J.A.S., and Turvey, M.T. (1980). On the concept of coordinative structures as dissipative structures. I. Theoretical lines of convergence. In: *Tutorials in motor behavior* (ed. G. E. Stelmach and J. Requin) pp. 3–47. North-Holland, Amsterdam.

Kuypers, H.G.J.M. (1962). Corticospinal connections: postnatal development in rhesus monkey. *Science* **138**, 678–80.

Lackner, J.R. (1973). The role of posture in adaptation to visual rearrangement. *Neuropsychologia* **11**, 33–44.

—— (1985). Human sensory-motor adaptation to the terrestrial force environment.

In: *Brain mechanisms and spatial vision* (ed. D. Ingle, M. Jeannerod, and D. Lee) pp. 175–209. M. Nijhoff, Dordrecht.

—— and Mather, J.A. (1981). Eye-hand tracking using afterimages. *Exp. Brain Res.* **44**, 138–42.

Lacquaniti, F. and Soechting, J.F. (1982). Coordination of arm and wrist motion during a reaching task. *J. Neurosci.* **2**, 399–408.

——, ——, and Terzuolo, C.A. (1982). Some factors pertinent to the organization and control of arm movements. *Brain Res.* **252**, 394–7.

Lamotte, R.H. and Acuna, C. (1978). Defects in accuracy of reaching after removal of posterior parietal cortex in monkeys. *Brain Res.* **139**, 309–26.

Langolf, G.D., Chaffin, D.B., and Foulke, J.A. (1976). An investigation of Fitts law using a wide range of movement amplitudes. *J. Mot. Behav.* **8**, 113–28.

Langworthy, O.R. (1933). Development of behavior patterns and myelinization of the nervous system in the human foetus and infant. *Contributions to embryology of the Carnegie Institution* Vol. 24, pp. 1–58.

Larish D.D., Volp, C.M., and Wallace, S.A. (1984). An empirical note on attaining a spatial target after distorting the initial conditions of movement via muscle vibration. *J. Mot. Behav.* **15**, 76–83.

Lashley, K.S. (1917). The accuracy of movement in the absence of excitation from the moving organ. *Am. J. Physiol.* **43**, 169–94.

—— (1951). The problem of serial order in behavior. In: *Cerebral mechanisms and behavior* (ed. L.A. Jeffress) pp. 112–36. Wiley, New York.

Lassek, A.M. (1953). Inactivation of voluntary motor function following rhizotomy. *J. Neuropath. Exp. Neurol.* **12**, 83–7.

—— and Moyer, E.K. (1953). An ontogenetic study of motor deficits following dorsal brachial rhizotomy. *J. Neurophysiol.* **16**, 247–51.

Laszlo, J. (1966). The performance of a simple motor task with kinaesthetic sense loss. *J. Exp. Psychol.* **18**, 1–8.

Latto, R. (1978). The effects of bilateral frontal eye-field, posterior parietal or superior colliculus lesions on visual search in the rhesus monkey. *Brain Res.* **146**, 35–50.

Laurutis, V.P. and Robinson, J.A. (1986). The vestibulo-ocular reflex during human saccadic eye movements. *J. Physiol.* **373**, 209–33.

LaVerne Morgan, C. (1978). Constancy of egocentric visual direction. *Percept. Psychophys.* **23**, 61–8.

Lawrence, D.G. and Hopkins, D.A. (1972). Development aspects of pyramidal control in the rhesus monkey. *Brain Res.* **40**, 117–18.

—— and Kuypers, H.G.J.M. (1968). The functional organization of the motor system in the monkey. I. The effects of bilateral pyramidal lesions. *Brain* **91**, 1–14.

Lee, D.N. (1976). A theory of visual control of braking based on information about time to collision. *Perception* **5**, 437–59.

—— and Reddisch, P.E. (1981). Plummeting gannets: a paradigm of ecological optics. *Nature* **293**, 293–4.

——, Young, D.S., Reddisch, P.E., Lough, S., and Clayton, T.M.H. (1983). Visual timing in hitting an accelerating ball. *Q. J. Exp. Psychol.* **35A**, 333–46.

Lee, R.G., and Tatton, W.G. (1975). Motor responses to sudden limb displacements in primates with specific CNS lesions and in human patients with motor system disorders. *Canad. J. Neurol. Sci.* **2**, 285–93.

Lee, W.A., and Kelso, J.A.S. (1979). Properties of slowly adapting joint receptors do not readily predict perception of limb position. *J. Hum. Mov. Stud.* **5**, 171–81.

Lenox, L.R., Lange, A.F., and Graham, K.R. (1970). Eye movement amplitudes in imagined pursuit of a pendulum with closed eyes. *Psychophysiology* **6**, 773–7.

Lestienne, F. (1979). Effects of inertial load and velocity on the braking process of voluntary limb movements. *Exp. Brain Res.* **35**, 407–18.

Levine, D.N., Kaufman, K.J., and Mohr, J.P. (1978). Inaccurate reaching associated with a superior parietal lobe tumor. *Neurology* **28**, 556–61.

Lewes, G.M. (1879). Motor feelings and muscular sense. *Brain* **1**, 14–28.

Liederman, J. and Kinsbourne, M. (1980). The mechanism of neonatal rightward turning bias. A sensory or motor asymmetry? *Infant behavior and development* **3**, 223–38.

Liu, C.N. and Chambers, W.W. (1971). A study of cerebellar dyskinesia in the bilaterally deafferented forelimbs of the monkey (*Macaca mulatta* and *Macaca speciosa*). *Acta Neurobiol. Exp.* **31**, 363–89.

Lloyd, A.J., and Caldwell, L.S. (1965). Accuracy of active and passive positioning of the leg on the basis of kinesthetic cues. *J. Comp. Physiol. Psychol.* **60**, 102–6.

Lockman, J.J., Ashmead, D.H., and Bushnell, E.W. (1984). The development of anticipatory hand orientation during infancy. *J. Exp. Child Psychol.* **37**, 176–86.

Lomas, J. and Kimura, D. (1976). Intrahemispheric interaction between speaking and sequential manual activity. *Neuropsychologia* **14**, 23–33.

Lough, S., Wing. A.M., Fraser, C., and Jenner, J.R. (1984). Measurement of recovery of function in the hemiparetic upper limb following stroke: a preliminary report. *Hum. Mov. Sci.* **3**, 247–56.

Lund, J.S., Downer, J.L. de C., and Lumley, J.S.P. (1970). Visual control of limb movement following section of optic chiasm and corpus callosum in the monkey. *Cortex* **6**, 323–46.

Luria, A.R. (1959). Disorders of 'simultaneous perception' in a case of bilateral occipito-parietal brain injury. *Brain* **82**, 437–49.

Lynch, J.C. (1980). The functional organization of posterior parietal association cortex. *Behav. Brain Sci.* **3**, 485–98.

—— and McLaren, J.W. (1979). Effects of lesions of parieto-occipital association cortex upon performance of oculomotor and attention tasks in monkeys. *Neurosci. Abstr.* **5**, 794.

——, Mountcastle, V.B., Talbot, W.H., and Yin, T.C.T. (1977). Parietal lobe mechanism for detecting visual attention. *J. Neurophysiol.* **40**, 362–89.

McCloskey, D.I. (1981). Corollary discharges: motor commands and perception. In: *Handbook of physiology, Section I: The nervous system*, Vol. 2: *Motor control* (ed. V.B. Brooks) Part 2, pp. 1415–47. American Physiological Society, Bethesda.

—— and Torda, T.A.G. (1975). Corollary motor discharges and kinaesthesia. *Brain Res.* **100**, 467–70.

——, Ebeling, P., and Goodwin, G.M. (1974). Estimation of weights and tensions and apparent involvement of a 'sense of effort'. *Exp. Neurol.* **42**, 220–32.

——, Cross, M.J., and Honner, R., and Potter, E.K. (1983a). Sensory effects of pulling or vibrating exposed tendons in man. *Brain* **106**, 21–37.

——, Gandevia, S., Potter, E.K., and Colebatch, J.G. (1983b). Muscle sense and effort. Motor commands and judgements about muscular contractions. In: *Motor control mechanisms in health and disease* (ed. J. E. Desmedt) pp. 151–167. Raven Press, New York.

McFie, J., Piercy, M.F., and Zangwill, O.L. (1950). Visual spatial agnosia associated with lesions of the right hemisphere. *Brain* **73**, 167–90.

McLaughlin, S.C. and Webster, R.G. (1967). Changes in straight-ahead eye position during adaptation to wedge prisms. *Percept. Psychophys.* **2**, 37–44.

McLeod, P. (1987). Visual reaction time and high-speed ball games. *Perception* **16**, 49–59.

MacNeilage P.F. (1970). Motor control of serial ordering of speech. *Psychol. Rev.* **77**, 182–96.

Mach, E. (1906). *The analysis of sensations.* Transl. from the 5th German edn. Dover, New York, 1959.

Manchester, D. (1985). The role of vision in grasp formation in prehension. Master Thesis, University of Oregon.

Mann, V.A., Hein, A., and Diamond, R. (1979). Localization of targets by strabismic subjects: contrasting patterns in constant and alternating suppressors. *Percept. Psychophys.* **25**, 29–34.

Marie, P., Bouttier, and Bailey, P. (1922). La planotopokinésie. Etude sur les erreurs d'exécution de certains mouvements dans leurs rapports avec la représentation spatiale. *Rev. neurol.* **29**, 505–12.

Marr, D. (1982). *Vision.* W. H. Freeman, San Francisco.

Marsden, C.D. Obeso, J.A., and Rothwell, J.C. (1983). The function of the antagonist muscle during fast limb movements in man. *J. Physiol.* **335**, 1–13.

——, Merton, P.A., Morton, H.B., Adam, J.E.R., and Hallett, M. (1978). Automatic and voluntary responses to muscle stretch in man. In: *Cerebral motor control in man: long loop mechanisms* (ed. J. E. Desmedt) pp. 167–77. *Progress in Clinical Neurophysiology.* Karger, Basel.

Marshall, J.C. (1984). Multiple perspectives on modularity. *Cognition* **17**, 209–42.

Marteniuk, R.G. (1978). The role of eye and head positions in slow movement execution. In: *Information processing in motor learning and control* (ed. G. E. Stelmach) pp. 267–88. Academic Press, New York.

——, Mackenzie, C.L., and Baba, D.M. (1984). Bimanual movement control: information processing and interaction effects. *Q. J. Exp. Psychol.* **36A**, 335–65.

——, Jeannerod, M., Athenes, S., and Dugas, C. (1987). Constraints on human arm movements trajectories. *Canad. J. Psychol.* (in press).

Mather, J.A. (1985). Some aspects of the motor organization of the oculomotor system. *J. Mot. Behav.* **17**, 373–83.

—— and Lackner, J.R. (1980). Visual tracking of active and passive movements of the hand. *Q. J. Exp. Psychol.* **32**, 307–15.

Matin, L. Eye movements and perceived visual direction. In: *Handbook of sensory physiology* (ed. D. Jameson and L. Hurvich) Vol. VII, pp. 331–80. Springer-Verlag, Berlin.

—— and Kibler, G. (1966). Acuity of visual perception of direction in the dark for various positions of the eye in the orbit. *Percept. Mot. Skills* **22**, 407–20.

—— and Pearce, D.G. (1965). Visual perception of direction for stimuli flashed during voluntary saccadic eye movements. *Science* **148**, 1485–8.

——, Pearce, D.G., Matin, E., and Kibler, G. (1966). Visual perception of direction. Roles of local sign, eye movements and ocular proprioception. *Vision Res.* **6**, 453–69.

——, Stevens, J.K., and Picoult, E. (1983). Perceptual consequences of experimental extraocular muscle paralysis. In: *Spatially oriented behavior* (ed. A. Hein and M. Jeannerod) pp. 243–62. Springer-Verlag, New York.

Matthews, P.B.C. (1959). The dependence of tension upon extension in the stretch-reflex of the soleus muscle of the decerebrate cat. *J. Physiol.* **147**, 521–46.

Matthews, P.B.C. (1982). Where does Sherrington's muscular sense originate? Muscle, joints, corollary discharges? *Ann. Rev. Neurosci.* **5**, 189–218.

Mauguière, F., Courjon, J., and Schott, B. (1983). Dissociation of early SEP components in unilateral traumatic section of the lower medulla. *Ann. Neurol.* **13**, 309–13.

Mays, L.E. and Sparks, D.L. (1980). Saccades are spatially, not retinocentrically, coded. *Science* **208**, 1163–5.

Megaw, E.D. (1974). Possible modification to a rapid on-going programmed manual response. *Brain Res.* **71**, 425–41.

—— and Armstrong, W. (1973). Individual and simultaneous tracking of a step input by the horizontal saccadic eye movement and manual control system. *J. Exp. Psychol.* **100**, 18–28.

Meinck, H.M., Benecke, R., Meyer, W., Höhne, J., and Conrad, B. (1984). Human ballistic finger flexion: uncoupling of the three-burst pattern. *Exp. Brain Res.* **55**, 127–33.

Merton, P.A. (1964). Human position sense and sense of effort. Soc. Exp. Biol., Symp. XVIII, *Homeostasis and feedback mechanisms*, pp. 387–400. Cambridge University Press, Cambridge.

Mesulam, M.M. and Geschwind, N. (1978). On the possible role of neocortex and its limbic connections in the process of attention and schizophrenia. Clinical cases of inattention in man and experimental anatomy in monkey. *J. Psychiat. Res.* **14**, 249–59.

Meyer, D.E., Smith, J.E.K., and Wright, C.E. (1982). Models for the speed and accuracy of aimed movements. *Psychol. Rev.* **89**, 449–82.

Meyer, D.L. and Bullock, T.H. (1977). The hypothesis of sense-organ-dependent tonus mechanisms: history of a concept. *Am. N.Y. Acad. Sci.* **290**, 3–17.

——, Abrams, R.A., Kornblum, S., Wright, C.E., and Smith, J.E.K. (Unpublished manuscript). Optimality in human motor performance: ideal control of rapid aimed movements.

Michel, F. (1971). Etude expérimentale de la vitesse du geste graphique. *Neuropsychologia* **9**, 1–13.

——, Jeannerod, M. and Devic, M. (1965). Trouble de l'orientation visuelle dans les trois dimensions de l'espace. *Cortex* **1**, 441–6.

Mikaelian, H. and Held, R. (1964). Two types of adaptation to an optically-rotated visual field. *Am. J. Psychol.* **77**, 257–63.

Miles, F.A. and Evarts, E.V. (1979). Concepts of motor organization. *Ann. Rev. Psychol.* **30**, 327–62.

Milner, A.D., Ockelford, E.M., and Dewar, W. (1977). Visuospatial performance following posterior parietal and lateral frontal lesions in stumptail macaques. *Cortex* **13**, 350–60.

Milner, T.E. (1986). Controlling velocity in rapid movements. *J. Mot. Behav.* **18**, 147–62.

Mishkin, M. and Ungerleider, L.G. (1982). Contribution of striate inputs to the visuospatial functions of parieto-preoccipital cortex in monkeys. *Behav. Brain Res.* **6**, 57–77.

——, Lewis, M.E., and Ungerleider, L.G. (1982). Equivalence of parieto-preoccipital subareas for visuospatial ability in monkeys. *Behav. Brain Res.* **6**, 41–56.

Mittelstaedt, H. (1983). A new solution to the problem of the subjective vertical. *Naturwissenschaften* **70**, 272–81.

Moberg, E. (1983). The role of cutaneous afferents in position sense, kinaesthesia and motor function of the hand. *Brain* **106**, 1–19.

Mohler, C.W. and Wurtz, R.H. (1977). Role of striate cortex and superior colliculus in visual guidance of saccadic eye movements in monkeys. *J. Neurophysiol.* **40**, 74–94.

Morasso, P. (1981). Spatial control of arm movements. *Exp. Brain Res.* **42**, 223–7.

Mott, F.W. and Sherrington, C.S. (1985). Experiments upon the influence of sensory nerves upon movement and nutrition of the limbs. *Proc. R. Soc.* **B57**, 481–8.

Motter, B.C. and Mountcastle, V.B. (1981). The functional properties of the light sensitive neurons of the posterior parietal cortex studied in waking monkeys. Foveal sparing and opponent vector orientation. *J. Neurosci.* **1**, 3–26.

Mountcastle, V.B., Lynch, J.C., Georgopoulos, A., Sakata, H., and Acuna, C. (1975). Posterior parietal association cortex of the monkey: command functions for operations within extra-personal space. *J. Neurophysiol* **38**, 871–908.

Muir, R.B. (1985). Small hand muscles in precision grip: a corticospinal prerogative? *Exp. Brain Res.* Suppl. 10, 155–74.

Muir, R.B. and Lemon, R.N. (1983). Corticospinal neurons with a special role in precison grip. *Brain Res.* **261**, 312–16.

Munk, H. (1909). Über die Folgen Sensibilitätsverlustes der Extremität fur deren Motilität. Kap. XIII in: *Über die Funktionen von Hirn und Rückenmark, Gesammelte Mitteilungen* pp. 247–85. Hirschwald, Berlin.

Newcombe, F. and Russell, W.R. (1969). Dissociated visual perceptual and spatial deficits in focal lesions of the right hemisphere. *J. Neurol. Neurosurg. Psychiat.* (1969) **32**, 73–81.

Newell, K.M. and Houk, J.C. (1983). Speed and accuracy of compensatory responses to limb disturbances. *J. Exp. Psychol. Hum. Percept. Perform.* **9**, 58–74.

Norman, D.A. and Shallice, T. (1980) Attention to action: willed and automatic control of behavior. Human Information Processing Technical Report No. 99. University of California, San Diego.

Oldfield, R.C., and Zangwill, O.L. (1942). Head's concept of the schema and its application on contemporary British psychology. *Br. J. Psychol.* **32**, 267–86 and **33**, 58–64.

Olson, C.R. (1980). Spatial localization in cats reared with strabismus. *J. Neurophysiol.* **43**, 792–806.

Orban, G.A., Kennedy, H., and Bullier, J. (1986). Velocity sensitivity and direction selectivity of neurons in areas V1 and V2 of the monkey: influence of eccentricity. *J. Neurophysiol.* **56**, 462–80.

——, ——, and Maes, H. (1981). Response to movement of neurons in areas 17 and 18 of the cat: velocity sensitivity. *J. Neurophysiol.* **45**, 1043–58.

——, Van Calenbergh, F., DeBruyn, B., and Maes, H. (1985). Velocity discrimination in central and peripheral visual field. *J. Opt. Soc. Am.* **A2**, 1836–47.

O'Regan, J.K. (1984). Retinal versus extraretinal influence in flash localization during saccadic eye movements in the presence of a visible background. *Percept. Psychophys.* **36**, 1–14.

Orem, J., Schlag-Rey, M., and Schlag, J. (1973). Unilateral visual neglect and thalamic intralaminar lesions in the cat. *Exp. Neurol.* **40**, 784–97.

Paillard, J. (1982). The contribution of peripheral and central vision to visually guided reaching. In: *Analysis of visual behavior* (ed. D. J. Ingle, M. A. Goodale, and R. J. W. Mansfield) pp. 367–85. MIT Press, Cambridge, Mass.

—— and Beaubaton, D. (1974). Problèmes posés par les contrôles moteurs ipsi-latéraux après déconnexion hémisphérique chez le singe. In: *Les syndromes de disconnexion calleuse chez l'homme* (ed. F. Michel and B. Schott) pp. 137–71. Hôpital Neurologique, Lyon.

—— and Brouchon M. (1968). Active and passive movements in the calibration of position sense. In: *The neuropsychology of spatially oriented behavior* (ed. S. J. Freedman) pp. 37–55. Dorsey Press, Homewood, Ill.

—— and —— (1974). A proprioceptive contribution to the spatial encoding of position cues for ballistic movements. *Brain Res.* **71**, 273–84.

——, Jordan, P., and Brouchon, M. (1981). Visual motion cues in prismatic adaptation: evidence for two separate and additive processes. *Acta Psychol* **48**, 253–70.

Palka, J. (1969). Discrimination between movements of eye and object by visual interneurones of crickets. *J. Exp. Biol.* **50**, 723–32.

Pandya, D.K., and Seltzer, B. (1982). Intrinsic connections and architectonics of posterior parietal cortex in the rhesus monkey. *J. Comp. Neurol.* **204**, 196–210.

Passingham, R., Perry, H., and Wilkinson, F. (1978). Failure to develop a precision grip in monkeys with unilateral neocortical lesions made in infancy. *Brain Res.* **145**, 410–14.

Paterson, A. and Zangwill, O.L. (1944). Disorders of visual space perception associated with lesions of the right cerebral hemisphere. *Brain* **67**, 331–58.

Peele, T.L. (1944). Acute and chronic parietal lobe ablations in monkeys. *J. Neurophysiol.* **7**, 269–86.

Pélisson, D., and Prablanc, C. (1986). Vestibulo-ocular reflex (VOR) induced by passive head rotation and goal-directed saccadic eye movements do not simply add in man. *Brain Res.* **380**, 397–400.

——, Prablanc, C., Goodale, M.A., and Jeannerod, M. (1986). Visual control of reaching movements without vision of the limb. II. Evidence for fast nonconscious processes correcting the trajectory of the hand to the final position of a double-step stimulus. *Exp. Brain Res.* **62**, 303–11.

Perenin, M.T. and Jeannerod, M. (1975). Residual vision in cortically blind hemifields. *Neuropsychologia* **13**, 1–7.

—— and —— (1978). Visual function within the hemianopic field following early cerebral hemidecortication in man. I. Spatial localization. *Neuropsychologia* **16**, 1–13.

—— and —— (1983). Are extrageniculostriate pathways nonfunctional in man? *Behav. Brain Sci.* **6**, 458–9.

—— and Vighetto, A. (1983). Optic ataxia: a specific disorder in visuomotor coordination. In: *Spatially oriented behavior.* (ed. A. Hein and M. Jeannerod) pp. 305–26. Springer-Verlag, New York.

——, Jeannerod, M., and Prablanc, C. (1977). Spatial localization with paralysed eye muscles. *Ophthalmologica* **175**, 206–14.

——, Vighetto, A., Mauguière, F., and Fischer, C. (1979). L'ataxie optique et son intérêt dans la coordination oeil-main. *Lyon méd.* **242**, 349–58.

Peters, W. and Wenborne, A.A. (1936). The time pattern of voluntary movements. *J. Psychol.* **26**, 388–406 and **27**, 60–73.

Petrides, M. and Iversen, S.D. (1979). Restricted posterior parietal lesions in the rhesus monkey and performance on visuo-spatial tasks. *Brain Res.* **161**, 63–77.

Pew, R.W. (1974). Human perceptual-motor performance. In: *Human information*

processing. Tutorials in performance and cognition (ed. B. H. Kautowitz) Erlbaum, Hillsdale, N.J.

Phillips, C.G. (1986). *Movements of the hand*. Liverpool University Press, Liverpool.

—— and Porter, R. (1977). *Corticospinal neurons. Their role in movement*. Academic Press, London.

——, Powell, T.P.S., and Wiesendanger, M. (1971). Projections from low-
threshold muscle afferents of hand and forearm to area 3a of baboon's cortex. *J.
Physiol.* **217**, 419–46.

Pierrot-Deseilligny, C., Gray, F., and Brunet, P. (1986). Infarcts of both inferior
parietal lobules with impairment of visually-guided eye movements, peripheral
visual inattention and optic ataxia. *Brain* **109**, 81–97.

Pirozzolo, F.J. and Rayner, K. (1980). Handedness, hemispheric specialization and
saccadic eye movement latencies. *Neuropsychologia* **18**, 225–9.

Peoppel, E. (1973). Letter to the editor. *Nature* **243**, 231.

——, Held, R., and Frost, D. (1973). Residual visual function after brain wounds
involving the central visual pathways in man. *Nature* **243**, 295–6.

Pohl, W. (1973). Dissociation of spatial discrimination deficits following frontal and
parietal lesions in monkeys. *J. Comp. Physiol. Psychol.* **82**, 227–39.

Polit, A. and Bizzi, E. (1978). Processes controlling arm movements in monkeys.
Science **201**, 1235–7.

—— and —— (1979). Characteristics of the motor programs underlying arm movements in monkeys. *J. Neurophysiol.* **42**, 183–94.

Porac, C. and Coren, S. (1986). Sighting dominance and egocentric localization.
Vision Res. **26**, 1709–13.

Poulton, E.C. (1974). *Tracking skill and manual control*. Academic Press, New
York.

—— (1980). Range effects and asymmetric transfer in studies of motor skills. In:
Psychology of motor behavior and sport (ed. C. M. Nadeau *et al.*) pp. 339–59.
Human Kinetic Publ., Champaign.

Prablanc, C. and Jeannerod, M. (1973). Continuous recording of hand position in
the study of complex visuo-motor tasks. *Neuropsychologia* **11**, 123–5.

—— and —— (1974). Latence et précision des saccades en fonction de l'intensité,
de la durée et de la position rétinienne d'un stimulus. *Rev. Electroenceph. Neurophysiol. Clin.* **4**, 484–8.

—— and —— (1975). Corrective saccades: dependence on retinal reafferent signals. *Vision Res.* **15**, 465–9.

——, ——, and Tzavaras, A. (1975a). Independent and interdependent processes
in prism adaptation. In: *Aspects of neural plasticity* (ed. F. Vital-Durand and
M. Jeannerod) pp. 139–52. INSERM, Paris.

——, Tzavaras, A., and Jeannerod, M. (1975b). Adaptation of the two arms to
opposite prism displacements. *Q. J. Exp. Psychol.* **27**, 667–71.

——, Massé, D., and Echallier, J.F. (1978). Error correcting mechanisms in large
saccades. *Vision Res.* **18**, 557–60.

——, Echallier, J.F., Komilis, E., and Jeannerod, M. (1979a). Optimal response of
eye and hand motor systems in pointing at a visual target. I. Spatio-temporal
characteristics of eye and hand movements and their relationships when varying
the amount of visual information. *Biol. Cybern.* **35**, 113–24.

——, Echallier, J.F., Komilis, E., and Jeannerod, M. (1979b). Optimal response of eye and hand motor systems in pointing at a visual target. II. Static and dynamic visual cues in the control of hand movements. *Biol. Cybern.* **35**, 183–7.

——, Pelisson, D., and Goodale, M.A. (1986). Visual control of reaching movements without vision of the limb. I. Role of retinal feedback of target position in guiding the hand. *Exp. Brain Res.* **62**, 293–302.

Provins, K.A. (1958). The effect of peripheral nerve block on the appreciation and execution of finger movements. *J. Physiol.* **143**, 55–67.

Putterman, A., Robert, A., and Bregman, A. (1969). Adaptation of the wrist to displacing prisms. *Psychon. Sci.* **16**, 79–80.

Pylyshyn, Z. (1980). Computational models and empirical constraints. *Behav. Brain Sci.* **1**, 93–128.

Ratcliff, G., and Davies-Jones, G.A.B. (1972). Defective visual localization in focal brain wounds. *Brain* **95**, 46–60.

Requin, J. (1985). Looking forward to moving soon. Ante factum selective processes in motor control. In: *Attention and performance* (ed. M. I. Posner and O. S. M. Marin) Vol. XI, pp. 147–67. Erlbaum, Hillsdale, N.J.

Richer, P. (1895). Du mode d'action des muscles antagonistes dans les mouvements très rapides, alternativement de sens inverse. *C.R. Soc. Biol. Paris* **47**, 171–4.

Robinson, D.A. (1964). The mechanics of human saccadic eye movements. *J. Physiol. (Lond.)* **174**, 245–64.

—— (1970). Oculomotor unit behavior in the monkey. *J. Neurophysiol.* **33**, 393–404.

—— (1975). Oculomotor control signals. In: *Basic mechanisms of ocular motility and their clinical implications* (ed. G. Lennerstrand and P. Bach-y-Rita) pp. 337–4. Pergamon Press, Oxford.

Robinson, D.L. and Wurtz, R.H. (1976). Use of an extra-retinal signal by monkey superior colliculus neurons to distinguish real from self-induced stimulus movement. *J. Neuropohysiol.* **39**, 852–70.

——, Goldberg, M.E., and Stanton, G.B. (1978). Parietal association cortex in the primate. Sensory mechanisms and behavioural modulation. *J. Neurophysiol.* **41**, 910–32.

Rogal, L., Reible, G., and Fischer, B. (1985). Reaction times of the eye and the hand of the monkey in a visual reach task. *Neurosci. Lett.* **58**, 127–32.

Roll, J.P., Bard, C., and Paillard, J. (1981). Rôle des mouvements céphalogyres sur la précision d'un ajustement visuomoteur chez l'homme. *J. Physiol. (Paris)* **77**, 44A.

Roll, R., Bard, C., and Paillard, J. (1986). Head orienting contributes to the directional accuracy of aiming at distant targets. *Hum. Mov. Sci.* **5**, 359–71.

Rondot, P. (1978). Le geste et son contrôle visuel. Ataxie visuomotrice. In: *Du contrôle moteur à l'organisation du geste* (ed. H. Hécaen and M. Jeannerod) Masson, Paris.

Rose, J.E. and Mountcastle, V.B. (1959). Touch and kinesthesis. In: *Handbook of physiology, Section I: Neurophysiology* (ed. J. Field) Vol. 1, pp. 387–429. American Physiological Society, Washington, D.C.

Rose, P.K. and Abrahams, V.C. (1975). The effect of passive eye movement on unit discharge in the superior colliculus of the cat. *Brain Res.* **97**, 95–106.

Rosenbaum, D.A. (1980). Human movement initiation: specification of arm, direction and extent. *J. Exp. Psychol. Gen.* **109**, 444–74.

Ross, H.E. and Bischof, K. (1981). Wundt's view on sensations of innervation: a review. *Perception* **10**, 319–29.

Rothwell, J.C., Traub, M.M., Day, B.L., Obeso, J.A., Thomas, P.K., and Marsden, C.D. (1982). Manual motor performance in a deafferented man. *Brain* **105**, 515–42.

Roucoux, A., Guitton, D., and Crommelinck, M. (1980). Stimulation of the superior colliculus in the alert cat. II. Eye and head movements evoked when the head is unrestrained. *Exp. Brain Res.* **39**, 75–85.

Roy, E.A. (1983). Manual performance asymmetries and motor control processes: Subject-generated changes in response parameters. *Hum-Mov. Sci.* **2**, 271–7.

Roy, E.A. and Elliott, D. (1986). Manual asymmetries in visually directed aiming. *Canad. J. Psychol.* **40**, 109–121.

Rubens, A.B. (1985). Caloric stimulation and unilateral visual neglect. *Neurology* **35**, 1019–24.

Sakata H., Shibutani, H., Ito, Y., and Tsurugai, K. (1986). Parietal cortical neurons responding to rotary movement of visual stimulus in space. *Exp. Brain Res.* **61**, 658–63.

Saltzman, E. (1979). Levels of sensorimotor representation. *J. Math. Psychol.* **20**, 91–163.

—— and Kelso, J.A.S. (In press.) Skilled actions: a task dynamic approach. *Psychol. Rev.*

Sanders, M.D., Warrington, E.K., Marshall, J., and Weiskrantz, L. (1974). 'Blindsight': vision in a field defect. *Lancet* (20 April) 707–8.

Sanes, J.N., and Jennings, V.A. (1984). Centrally programmed patterns of muscle activity in voluntary motor behavior of humans. *Exp. Brain Res.* **54**, 23–32.

——, Mauritz, K.H., Dalakas, M.C., and Evarts, E.V. (1985). Motor control in humans with large-fiber sensory neuropathy. *Hum. Neurobiol.* **4**, 101–14.

Schaeffer, A.A. (1928). Spiral movement in man. *J. Morphol. Physiol.* **45**, 293–398.

Schaffer, L.H. (1981). Performances of Chopin, Bach, and Bartoli: studies in motor programming. *Cognitive Psychol.* **13**, 326–76.

Schilder, P. (1935). *The image and appearance of the human body*. Routledge & Kegan Paul, London.

Schmidt, R.A. (1975). A schema theory of discrete motor skill learning. *Psychol. Rev.* **82**, 225–60.

——, Zelaznik, H., and Frank, J.S. (1978). Sources of inaccuracy in rapid movement. In: *Information processing in motor control and learning* (ed. G. E. Stelmach) pp. 183–203. Academic Press, New York.

——, ——, Hawkins, B., Franck, J.S., and Quinn, J.T. (1979). Motor-output variability: a theory for the accuracy of rapid motor acts. *Psychol. Rev.* **86**, 415–51.

Schneider, G.E. (1967). Contrasting visuo-motor functions of tectum and cortex in the golden hamster. *Psychol. Forsch.* **31**, 52–62.

Schott, B., Jeannerod, M. and Zahin, M.Z. (1966). L'agnosie spatiale unilatérale: perturbation en secteur des mécanismes d'exploration et de fixation du regard. *J. méd. Lyon* **47**, 169–95.

Schutz, R.W., and Roy, E.A. (1973). Absolute error: the devil in disguise. *J. Mot. Behav.* **5**, 141–53.

Schwartz, D.W.F. and Tomlinson, R.D. (1977). Neuronal responses to eye muscle stretch in cerebellar lobule VI of the cat. *Exp. Brain Res.* **27**, 101–11.

Sherrington, C.S. (1893). Further experimental note on the correlation of action of antagonistic muscles. *Proc. R. Soc.* **B53**, 407–20.

—— (1894). On the anatomical constitution of nerves of skeletal muscles, with remarks on recurrent fibers in the ventral spinal nerve-root. *J. Physiol.* **17**, 211–58.

—— (1897). Further note on the sensory nerves of muscles. *Proc. R. Soc.* **B61**, 247–9.

—— (1898). Further note on the sensory nerves of muscles. *Proc. R. Soc.* **B62**, 120–1.

—— (1900). The muscular sense. In: *Textbook of physiology* (ed. E. A. Schäffer) Vol. 2, pp. 1002–25. Pentland, London.

Siebeck, R. (1954). Wahrnehmungsstörung und Störungswahrnehmung bei Augenmuskellähmungen. *A. Graefes Arch. klin. exp. Ophthal.* **155**, 26–34.

Silberpfennig, J. (1941). Contributions to problem of eye-movements. IV. Disturbances of ocular movements with pseudo-hemianopsia in frontal lobes tumors. *Confin. Neurol. (Basel)* **4**, 1–13.

Skavenski, A.A. (1971). Extraretinal correction and memory for target position. *Vision Res.* **11**, 743–6.

—— (1972). Inflow as a source of extraretinal eye position information. *Vision Res.* **12**, 221–9.

—— (1976). The nature and role of extraretinal eye-position information in visual localization. In: *Eye movements and psychological processes* (ed. R. A. Monty and J. W. Senders) pp. 277–87. Erlbaum, Hillsdale, N.J.

—— and Hansen, R.M. (1978). Role of eye position information in visual space perception. In: *Eye movements and the higher psychological functions* (ed. J. Senders, D. Fisher, and R. Monty) pp. 15–34. Erlbaum, New York.

—— and Steinman, R.M. (1970). Control of eye position in the dark. *Vision Res.* **10**, 193–203.

Slack, C.W. (1953). Some characteristics of the 'range effect'. *J. Exp. Psychol.* **46**, 76–80.

Smith, W.M. and Bowen, K.F. (1980). The effects of delayed and displaced visual feedback on motor control. *J. Mot. Behav.* **12**, 91–101.

Smith, W.M., McCrary, J.W., and Smith, K.U. (1960). Delayed visual feedback and behavior. *Science* **132**, 1013–14.

Soechting, J.F. (1984). Effect of target size on spatial and temporal characteristics of a pointing movement in man. *Exp. Brain Res.* **54**, 121–32.

—— and Lacquaniti, F. (1981). Invariant characteristics of a pointing movement in man. *J. Neurosci.* **1**, 710–20.

Sparks, D.L. and Mays, L.E. (1983). Role of the monkey superior colliculus in the spatial localization of saccade targets. In: *Spatially oriented behavior* (ed. A. Hein and M. Jeannerod) pp. 63–85. Springer-Verlag, New York.

Sperry, R.W. (1943). Effect of 180° rotation of the retinal field in visuomotor coordination. *J. Exp. Zool.* **92**, 263–79.

—— (1950). Neural basis of the spontaneous optokinetic response produced by visual inversion. *J. Comp. Physiol. Psychol.* **43**, 482–9.

Sprague, J.M. (1966). Interactions of cortex and superior colliculus in mediation of visually guided behavior in the cat. *Science* **152**, 1544–7.

—— and Meikle, T.H. (1965). The role of superior colliculus in visually guided behavior. *Exp. Neurol.* **11**, 115–46.

Stark, L. (1968). *Neurological control systems. Studies in bioengineering.* Plenum Press, New York.

Stein, J. (1978). Long-loop motor control in monkeys. The effects of transient cooling of parietal cortex and of cerebellar nuclei during tracking tasks. In: *Cerebral motor control in man: long-loop mechanisms* (ed. J. E. Desmedt) pp. 107–22. Karger, Basel.

Steinbach, M.J. and Held, R. (1968). Eye tracking of observer-generated target movements. *Science* **161**, 187–8.

—— and Smith, D.R. (1981). Spatial localization after strabismus surgery: evidence for inflow. *Science* **213**, 1407–9.

Stetson, R.H., and Bouman, H.D. (1935). The coordination of simple skilled movements. *Arch. Neerland. Physiol.* **20**, 179–254.

Stevens, J.K., Emerson, R.C., Gerstein, G.L., Kallos, T., Neufield, G.R., Nichols, C.W., and Rosenquist, A.C. (1976). Paralysis of the awake human: visual perceptions. *Vision Res.* **16**, 93–8.

Takemori, S., Uchigata, M., and Ishikawa, M. (1979). Eye movements in cerebral lesions. *Neurosci. Lett.*, Suppl. 2, S30.

Taub, E. (1976). Motor behavior following deafferentation in the developing and motorically immature monkey. In: *Neural control of locomotion* (ed. R. M. Herman, S. Grillner, D. G. Stein, and D. G. Stuart) pp. 675–705. Plenum Press, New York.

—— and Berman, A.J. (1968). Movement and learning in the absence of sensory feedback. In: *The neurophysiology of spatially oriented behavior* (ed. S. J. Freedman) pp. 173–92. Dorsey Press, Homewood.

——, Perrella, P., and Barro, G. (1973). Behavioral development after forelimb deafferentation on day of birth in monkeys with and without blinding. *Science* **181**, 959–60.

——, Goldberg, I.A., and Taub, P. (1975). Deafferentation in monkeys: pointing at a target without visual feedback. *Exp. Neurol.* **46**, 178–86.

Taylor, C.R., Heglund, N.C., McMathon, T.A., and Looney, T.R. (1980). Energetic cost of generating muscular force during running. *J. Exp. Biol.* **86**, 9–18.

Terzuolo, C.A., Soechting, J.F., and Viviani, P. (1973). Studies on the control of some simple motor tasks. I. Relations between parameters of movements and EMG activity. *Brain Res.* **58**, 212–16.

——, ——, and Ranish, N.A. (1974). Studies on the control of some simple motor tasks. V. Changes in motor output following dorsal root section in squirrel monkey. *Brain Res.* **70**, 521–6.

Teuber, H.L. (1960). Perception. In: *Handbook of physiology, Section I, Neurophysiology* (ed. J. Field, H. W. Magoun, and V. E. Hall) pp. 89–121. American Physiological Society, Washington, D.C.

Thomson, J.A. (1980). How do we use visual information to control locomotion? *Trends Neurosci.* **3**, 247–50.

Todor, J.I. and Cisneros, J. (1985). Accommodation to increased accuracy demands by the right and left hands. *J. Mot. Behav.* **17**, 355–72.

Todor, J.I., Kyprie, P.M., and Price, H.L. (1982). Lateral asymmetries in arm, wrist and finger movements. *Cortex* **18**, 515–23.

Tomlinson, R.D. and Bahra, P.S. (1986). Combined eye-head gaze shifts in the primate. I. Metrics. *J. Neurophysiol.* **56**, 1542.

Tower, S.S. (1940). Pyramidal lesion in the monkey. *Brain* **63**, 36–90.

Trevarthen, C.B. (1965). Functional interactions between the cerebral hemispheres in the monkey. In: *Functions of the corpus callosum* (ed. E. G. Ettlinger) pp. 24–40. Ciba Foundation, Churchill, London.

—— (1968). Two mechanisms of vision in primates. *Pyschol. Forsch.* **31**, 299–337.

—— (1982). Basic patterns of psychogenetic change in infancy. In: *Dips in learning* (ed. T. G. Bever) pp. 7–46. Erlbaum, Hillsdale, N.J.

—— (1984). How control of movement develops. In: *Human motor actions. Bernstein reassessed* (ed. H.T.A. Whiting) pp. 223–61. North-Holland, Amsterdam.

—— and Sperry, R.W. (1973). Perceptual unity of the ambient visual field in human commissurotomy patients. *Brain* **96**, 547–70.

Twitchell, T.E. (1954). Sensory factors in purposive movements. *J. Neurophysiol.* **17**, 239–52.

—— (1970). Reflex mechanisms and the development of prehension. In: *Mechanisms of motor skill development* (ed. H. Connolly) pp. 25–38. Academic Press, London.

Tyler, C.W. and Torres, J. (1972). Frequency response characteristics for sinusoidal movement in the fovea and periphery. *Percept. Psychophys.* **12**, 232–6.

Tzavaras, A. and Masure M.C. (1975). Aspects différents de l'ataxie optique selon la latéralisation hémisphérique de la lésion. *Lyon méd.* **236**, 673–83.

Uemura, T., Arai, Y., and Shimazaki, C. (1980). Eye-head coordination during lateral gaze in normal subjects. *Acta oto-laryng. (Stockholm)* **90**, 191–8.

Valenstein, E., Van den Abell, T., Watson, R.T., and Heilman, K.M. (1982). Non-sensory neglect from parieto-temporal lesions in monkeys. *Neurology* **32**, 1198–201.

Vallar, G. and Perani, D. (1986). The anatomy of unilateral neglect after right-hemisphere stroke lesions. A clinical/CT scan correlation study in man. *Neuropsychologia* **24**, 609–22.

Van der Staak, C. (1975). Intra- and interhemispheric visual-motor control of human arm movements. *Neuropsychologia* **13**, 439–48.

Vanni-Mercier, G. and Magnin, M. (1982). Retinotopic organization of extra-retinal saccade-related input to the visual cortex in the cat. *Exp. Brain Res.* **46**, 368–76.

Van Opstal, A.J., Van Gisbergen, J.A.M., and Eggermont, J.J. (1985). Reconstruction of neural control signals for saccades based on an inverse method. *Vision Res.* **25**, 789–801.

Ventre, J. (1985). Cortical control of oculomotor functions. II. Vestibulo-ocular reflex and visual-vestibular interaction. *Behav. Brain Res.* **1**, 221–34.

—— and Faugier-Grimaud, S. (1986). Effects of posterior parietal lesions (area 7) on VOR in monkeys. *Exp. Brain Res.* **62**, 654–8.

——, Flandrin, J.M., and Jeannerod, M. (1984). In search for the egocentric reference. A neurophysiological hypothesis. *Neuropsychologia* **22**, (6), 797–806.

Vighetto, A. (1980). Etude neuropsychologique et psychophysique de l'ataxie optique. Thèse de Médecine, Lyon.

Vince, M.A. (1948). Corrective movements in a pursuit task. *Q. J. Exp. Psychol.* **1**, 85–106.

Viviani, P., and McCollum, G. (1983). The relation between linear extent and velocity in drawing movements. *Neuroscience* **10**, 211–18.

—— and Terzuolo, C. (1980). Space-time invariance in learned motor skills. In: *Tutorials in motor behavior* (ed. G. E. Stelmach and J. Requin) pp. 525–533. North-Holland, Amsterdam.

—— and —— (1982). Trajectory determines movement dynamics. *Neuroscience* **7**, 431–7.

Vogt. C. and Vogt, O. (1919). Allgemeine Ergebnisse unserer Hirnforschung. *J. Psychol. Neurol.* **25**, 279–462.

von Bonin, G. and Bailey, P. (1947). *The neocortex of Macaca mulatta*. University of Illinois Press, Urbana.

von Graefe, A. (1870). *Les paralysies des muscles moteurs de l'oeil*. Transl. from German by A. Sichel. Delahaye, Paris.

von Helmholtz, H. (1866). *Handbuch des physiologischen Optik*. Vos, Leipzig. French transl., Masson, Paris, 1867; English transl., Optical Society of America, 1925.

von Hofsten, C. (1979). Development of visually-directed reaching: the approach phase. *J. Hum. Mov. Stud.* **5**, 160–78.

—— (1982). Eye-hand coordination in the newborn. *Develop. Psychol.* **18**, 450–61.

—— and Fazel-Zandy, S. (1984). Development of visually-guided hand orientation in reaching. *J. Exp. Psychol.* **38**, 208–19.

von Holst, E. and Mittelstaedt, H. (1950). Das Reafferenzprinzip. Wechselwirkungen zwischen Zentralnervensystem und Peripherie. *Naturwissenschaften* **37**, 464–76.

von Noorden, G.K., Awaya, S., and Romano, P.E. (1971). Past-pointing in paralytic strabismus. *Am. J. Ophthal.* **71**, 27–33.

Wachholder, K. (1928). Willkürliche Haltung und Bewegung insbesondere im Lichte electrophsiologischer Untersuchungen. *Ergebn. Physiol.* **24**, 568–775.

—— and Altenburger, H. (1926). Beiträge zur Physiologie der willkürlichen Bewegung. X Mitteilung. Einzelbewegungen. *Pflügers Arch. ges. Physiol.* **214**, 642–661.

Wade, N.J. (1978). Sir Charles Bell on visual direction. *Perception* **7**, 359–62.

Wadman, W.J., Denier Van der Gon, J.J., Geuze, R.H., and Mol. C.R. (1979). Control of fast goal-directed arm-movements. *J. Hum. Mov. Stud.* **5**, 3–17.

Wallace, S.A. and Newell, K.M. (1983). Visual control of discrete aiming movements. *Q. J. Exp. Psychol.* **35A**, 311–21.

Waller, A.D. (1891). The sense of effort. An objective study. *Brain* **14**, 179–249.

Warabi, T. (1977). The reaction time of eye–head coordination in man. *Neurosci. Lett.* **6**, 47–51.

Waters, P. and Strick, P.L. (1981). Influence of 'strategy' on muscle activity during ballistic movements. *Brain Res.* **207**, 189–94.

Watson, R.T., Valenstein, E., Day, A.L., and Heilman, K.M. (1984). The effects of corpus callosum lesions on unilateral neglect in monkeys. *Neurology* **34**, 812–15.

Weiskrantz L., Warrington, E.R., Sanders, M.D., and Marshall, J. (1974). Visual capacity in the hemianopic field following a restricted occipital ablation. *Brain* **97**, 709–28.

Welford, A.T. (1968). *Fundamentals of skills*. Methuen, London.

Werner, H. and Wapner, S. (1952). Toward a generalized theory of perception. *Psychol. Rev.* **59**, 324–38.

Werner, H. Wapner, S., and Bruell, J.H. (1953). Experiments on sensory-tonic field theory of perception: VI. Effect of position of head, eyes and of object on position of the apparent median plane. *J. Exp. Psychol.* **46**, 293–9.

Whiting, H.T.A. (Editor) (1984). *Human motor actions. Bernstein reassessed*. *Advances in Psychology*, Vol. 17. North-Holland, Amsterdam.

——, Gill, E.B., and Stephenson, J.M. (1970). Critical time intervals for taking in flight information in a ball-catching task. *Ergonomics* **13**, 265–72.

Wiersma, C.A.G. and Yamaguchi, T. (1967). Integration of visual stimuli by the crayfish central nervous system. *J. Exp. Biol.* **47**, 409–31.

Wild, M.J. and Zeigler, H.P. (1980). Central representation and somatotopic organization of jaw muscles within the facial and trigeminal nuclei of the pigeon (Columba Livia). *J. Comp. Neurol.* **192**, 175–201.

Wing, A.M. and Fraser, C. (1983). The contribution of the thumb to reaching movements. *Q. J. Exp. Psychol.* **35A**, 297–309.

——, Turton, A., and Fraser, C. (1986). Grasp size and accuracy of approach in reaching. *J. Mot. Behav.* **18**, 245–60.

Woodworth, R.S. (1899). The accuracy of voluntary movements. *Psychol. Rev. Monogr.* Suppl. 3.

Woolsey, C.N., Gorska, T., Wetzel, A., Erickson, T.C., Earls, F.J., and Allman J.M. (1972). Complete unilateral section of the pyramidal tract at the medullary level in *Macaca mulatta*. *Brain Res.* **40**, 119–23.

Wundt, W. (1892). *Vorlesungen über die menschen und Tierseele*, 2nd edn. Voss, Hamburg.

Yakovlev, P.I. and Lecours, A.R. (1967). The myelogenetic cycles of regional maturation of the brain. In: *Regional developmnt of the brain in early life* (ed. A. Minkowski) pp. 3–70. Blackwell, Oxford.

Yin, T.C.T. and Mountcastle, V.B. (1977). Visual input to the visuomotor mechanisms of the monkey's parietal lobe. *Science* **197**, 1381–3.

Young, L.R. and Stark, L. (1963a). Variable feedback experiments testing a sampled-data model for eye tracking movement. *IEEE Trans. Hum. Factors Electro.* **HFE4**, 38–51.

—— and —— (1963b). A discrete model for eye tracking movements. *IEEE Trans.* **MIL7**, 113–15.

Zangemeister, W.M. and Stark, L. (1982). Gaze latency: variable interactions of head and eye latency. *Exp. Neurol.* **75**, 389–406.

——, Jones, A., and Stark, L. (1981). Dynamics of head movement trajectories: main sequence relationship. *Exp. Neurol.* **71**, 76–91.

Zelaznik, H.N., Hawkins, B., and Kisselburg, K. (1983). Rapid visual feedback processing in single-aiming movements. *J. Mot. Behav.* **15**, 217–36.

——, Schmidt, R.A., and Gielen, S.C.A.M. (1986). Kinematic properties of rapid aimed hand movements. *J. Mot. Behav.* **18**, 353–72.

Zihl, J. (1980). Blindsight: improvement of visually guided eye movements by systematic practice in patients with cerebral blindness. *Neuropsychologia* **18**, 71–7.

—— and Werth, R. (1984). Contributions to the study of 'blindsight'. I. Can straylight account for saccadic localization in patients with postgeniculate field defects? *Neuropsychologia* **22**, 1–11.

—— and —— (1984). Contributions to the study of 'blindsight'. II. The role of specific practice for saccadic localization in patients with postgeniculate visual field defects. *Neuropsychologia* **22**, 13–22.

Author index

Subject index

proximo-distal organization of
    movements 55–80
pyramidal tract 8, 68–71, 75, 189
pyramidotomy 69–71

range effect 110, 127, 128
reconstruction of target position 47, 49, 51,
    132–5, 196–9
recovery (of function) 70, 71
reference (for spatial behaviour) 5, 51, 103,
    168, 209, 221
representation (of goal of movement) 1–3,
    22, 28, 39, 129, 131, 196–202
representation (of space) 170, 214
residual vision (following lesions of visual
    cortex) 125–30
rhizotomy 6, 23–5, 173
right–left asymmetry (in movements) 45, 46,
    77, 78, 82, 99, 161

saccades, *see* eye movements
schema (motor, perceptual) 1, 2, 30, 58, 158
sensation of effort 172, 173, 188, 189, 191;
    see also centralist theory
sensation of heaviness 188, 189, 190
sensation of innervation, *see* sensation of
    effort
shaping (of the hand) 56, 68, 75
skin receptors (contribution in position
    sense) 185, 186
spasticity 71–3
spatial map 37, 39, 131, 132–6, 170, 192–9,
    241–4
spatial neglect 167–70, 210, 214, 215, 221,
    222, 226–8, 238
speech movements 22, 28, 32, 33, 200, 201
speed–accuracy tradeoff 85, 86, 90–6, 110
split-brain 75, 76

stability (of vision) 137, 140, 141, 148, 172,
    173
strabismus, *see* eye deviation
straight ahead direction (perception of) 147,
    152, 158–65, 167, 217–21
superior colliculus 44, 125, 133, 134, 135,
    137, 166–8
synergies 28–30, 39, 43, 46, 47, 49, 157

tendon pulling (subjective effects of) 184,
    185, 186
time to collision 101, 102
trajectory formation 26, 34, 36, 37, 39, 60
transfer (of adaptation) 52–6
triphasic EMG pattern, *see*
    electromyographic activity during
    movement
typing 29, 33, 34

variability (of motor impulses) 14, 18, 90,
    95, 119
vestibular responses (asymmetry in) 168–70,
    215
vibration of muscles (effects of) 26, 151–4,
    183, 184
visual direction (perception of) 144, 147,
    161, 162, 164, 165
visual feedback (delayed) 113, 114
visual feedback (delay of) 87, 89, 93, 95–105
visual feedback (selective exclusion of) 99,
    100, 105, 107–20, 175, 192–200,
    203–8, 217–19, 223–6
visual feedback (terminal) 100, 120, 123
visuoconstructive disorders 214, 215

writing 33–6